KU-282-924

Jupiter

The Giant Planet

♃

Reta Beebe

Smithsonian Institution Press
Washington
London

To My Mother and Father
Elsie E. McBride and Thomas H. Jenkins

Copy and Production Editor: Diane Amussen
Designer: Alan Carter
Library of Congress Cataloguing-in-Publication Data
Beebe, Reta.
Jupiter / Reta Beebe.
p. cm.—(Smithsonian Library of the solar system)
Includes bibliographical references and index.
ISBN 1-56098-417-1
1. Jupiter (Planet) I. Title. II. Series.
QB661.B44 1994
523.4′5—dc20
British Library Cataloguing-in-Publication Data is available
Manufactured in the United States of America
02 01 00 99 98 97 96 95 5 4 3 2 1
∞The paper used in this publication meets the minimum requirements of the American National Standard for permanence of Paper for Printed Library Materials Z39.48–1984.

Contents

Introduction

Throughout the night Jupiter appears as a bright starlike object, moving eastward relative to the background stars at a rate of about thirty degrees per year. Even with the earliest telescopes this curious object could be resolved into a pearly banded disk, and although observations span more than 380 years, this giant planet continues to fascinate and challenge amateur and professional observers.

Since the advent of space probes the scope of the available data has expanded to such an extent that researchers are drawn from a broad spectrum of scientific and technical fields. Astronomers, chemists, engineers, geologists, meteorologists, and plasma physicists collaborate with computer and telemetry experts to interpret the data. Because the jovian system is complex and its consideration requires many areas of expertise, any survey of our knowledge of it is a formidable task.

Jupiter is a rapidly rotating gas-giant located five times farther from the sun than our earth and receiving effective solar heating equal to 4 percent of that incident on the earth's atmosphere. It has an atmospheric environment so frigid that ammonia ice crystals form an opaque cloud which blocks our line of sight into the atmosphere. Below this cloud deck pressures increase rapidly with depth, compressing the gases more and more until, deep below, the compressed material behaves more like a liquid. Nevertheless, a penetrating probe would not encounter a liquid ocean or a solid surface, but would instead eventually be crushed as the pressure increased.

Formed in the outer solar system, the planet and its surrounding environment differ greatly from ours. Even though Jupiter is

318 times as massive as the earth, its average density is less than one-fourth that of the earth. This fact and spectroscopic observations indicate that the chemical composition is similar to that of the sun—consisting mostly of hydrogen and helium. Even though the planet is so much more massive than the earth, it is still only one-thousandth the mass of the sun. Furthermore, infrared observations have revealed that Jupiter is radiating more than one and a half times as much heat as it gains from the sunlight it absorbs. Thus Jupiter plays two roles: the dominant sibling among the planets and a substellar companion of the sun. In either role, Jupiter is looked upon as a laboratory within which the extremes of any physical model for understanding the past and present conditions of planet and star formation may be tested.

The jovian system has the characteristics of a miniature solar system (the distance between the sun and Jupiter is about 1000 times greater than that between Jupiter and the innermost large satellite, Io). Jupiter's "planetary system" consists of an equatorial ring of debris and 16 moons which include the 4 small inner satellites, the 4 large Galilean satellites, and two sets of 4 small satellites located about 30 to 60 times farther from the planet than Io. The 12 inner satellites revolve around the planet in the same sense that the planet rotates on its axis, and of these, the 8 innermost satellites orbit nearly in the equatorial plane. The 4 outermost satellites appear to be leftover debris formed by a collisional process and revolve about the planet in reverse relative to Jupiter's rotation. Like their 4 nearest siblings, their orbits are inclined to Jupiter's equator. Overall, the degree of order that we see in this system suggests that although episodic events may have occurred, at least the innermost set of bodies formed as an original complement and that long-term processes throughout the ages have molded the system into its present state.

The fact that Jupiter is transferring heat from its interior and is rotating rapidly allows it to sustain a magnetic field. At the level of the visible cloud deck, this field is 20 to 30 times as strong as the earth's. Charged particles streaming out from the sun do not bombard Jupiter's cloud tops; instead they are captured and forced to co-rotate with its magnetic field. As the planet rotates, these particles are swept past the inner satellites, creating a system far more complex than our own. Again, this planetary giant provides us with a unique arena for testing our understanding of the principles that underlie the generation and maintenance of our own magnetic field, our shield from the high-energy particles with which the sun bombards us.

In this book I have divided the complex jovian system into logical units: the atmosphere and interior of the planet, the satellites and ring system, and the magnetic field. In each case I have attempted to present a brief review of the history and enough detail for the reader to appreciate our current state of knowledge. In the interest of readability I have limited the detailed description of Jupiter's cloud structure and of the extensive mapping and nomenclature assigned to surface features on the jovian satellites.

Finally, in the last part we look to the future and speculate about what we may learn about the whole jovian system from the Galileo Mission and from the collision with Comet Shoemaker-Levy 9 in July 1994.

This attempt to summarize the broad scope of our current understanding of Jupiter's complex system has often been frustrating. Although I have spent a major portion of the last two decades studying this planet and have participated in all six Voyager encounters, the production of this book has been severely handicapped by the level of involvement that I could commit to the project. When I have succeeded in conveying my intent, much of the credit goes to those who read and criticized my efforts. If I have failed to cite significant contributors or omitted favorite topics, I beg their pardon and yours. I hope that my efforts will stimulate your interest and enhance your appreciation of the order and complexity of the solar system. I hope also that when you have considered the scope of this immense system and the degree to which we have only begun to tap its secrets, you will humbly add what you have gained to a better understanding of our small place in the universe.

24 PART I

A HISTORICAL REVIEW

♃ Chapter 1

From Mythology to Robotic Exploration

Jupiter is the largest of all planets and its system of sixteen satellites is impressive. But even before these things were known this bright, moving object had been given the Roman name for the lord of the sky. Because Jupiter revolves about the sun in an orbit with a radius more than five times that of the earth, it can be seen in our night sky for most of the year. Each year Jupiter's orbital motion causes it to appear to shift about 30° eastward against the background stars. This eastward motion creates a twelve-year cycle, so that each year when Jupiter is visible on your meridian (a line connecting the north point on your horizon, the point directly over your head, and the south point on your horizon) at midnight, it can be associated with a different sign of the zodiac. The predictability of this cycle and the brilliance of the planet relative to nearby stars suggests control and dominance and implies how this planet came to be named for the Roman god Jupiter.

As more became known about the physical nature of the planets, Jupiter lived up to its name. Not only is this planet more than three times as massive as Saturn, the next largest planet, but also it possesses four of the largest satellites, or moons, in our solar system (see appendix 1). These large satellites are readily seen with ordinary binoculars and are collectively referred to as the Galilean satellites, in honor of their discoverer, Galileo Galilei (1564–1642). The common names of these four are Io, Europa, Ganymede, and Callisto.

According to Roman legend, Io, Europa, and Callisto were beautiful females who were victims of Jupiter's lust. When Jupiter's interest in Io enraged Juno, Jupiter's wife and sister, Jupiter changed Io into a heifer to conceal his infidelity. This did not fool Juno or deter her from seeking revenge. She sent a gadfly to harass Io for eternity.

Callisto's legend is strikingly parallel. This time it was Juno who changed the lovely arcadian nymph into a bear to spite Jupiter. When Callisto's son encountered his mother during a hunting trip, Jupiter cast a spell and saved Callisto by placing her in the northern sky. Later he changed the son into a bear cub and placed him next to his mother. They are known as the constellations of Ursa Major and Ursa Minor, the Great Bear and Lesser Bear.

In still another legend the ever-lustful Jupiter appeared at the seashore as a great white bull and kidnapped Europa, daughter of Agenor, king of Phoenicia. Jupiter carried Europa off to the isle of Crete, where she bore him four sons. According to some legends she was also the mother of the Minotaur.

In contrast, the name of Ganymede comes from Greek sources. Why? We do not know. Ganymede was a Trojan lad whose beauty was so pleasing to Zeus (Jupiter's Greek counterpart) that he sent an eagle to bring the boy to Mount Olympus, where he became Zeus' cupbearer. Homer called Ganymede the most beautiful of mortal men. This brief review illustrates the internationally acceptable tradition of naming the satellites of a planet for associated mythological characters.

Even in small telescopes Jupiter appears as a banded disk; the four bright accompanying satellites look like small stars in orbit about it, revolving parallel to the bands. Careful inspection reveals distinct clouds that appear on the western edge of the disk and move eastward; they indicate that the planet rotates rapidly, completing one full rotation in less than ten hours (plate 1*a*). Jupiter's easy visibility has made it a favorite target of astronomers and amateur observers over the years.

Changes in color and texture of the bands led early observers to suspect that these were a changing cloudscape overlying the planet's surface. Their desire to know more about the planet's surface caused them to pay special attention to the appearance of darker areas, which they thought were breaks in the brighter clouds. Their efforts proved futile, and by the time spacecraft exploration began, astronomers were convinced that the outer regions of this planet contained layer on layer of frigid clouds and that a solid earthlike surface did not exist.

Although Jupiter is 318 times more massive than the earth, its average density is less than one-fourth the earth's. Such low average density means that the amount of dense rocky and metallic material that can be present within the planet's cloud-shrouded interior is severely limited and that, like the sun, Jupiter must be composed mainly of hydrogen and helium. Because this planet is so unlike earth, it presents a major challenge to investigators.

Within the past two decades the National Aeronautical and Space Agency (NASA) has sent Pioneer and Voyager spacecraft to Jupiter for closeup views of it and its moons. In all four of these flybys much of the data collected concerned only the outer few hundred kilometers of the 71,500-km radius (equivalent to 11.3 earth radii). Even though the clouds have frustrated efforts of investigators to probe deeper, the knowledge gained from these missions indicates that the gaseous atmosphere must extend for tens of thousands of kilometers below the visible cloud layer. The dense core is buried so deeply below this huge obscuring atmosphere that it is impossible to reach. Investigators are forced to resort to indirect methods to learn more about the planet's interior.

Jupiter has had an important historical role in the astronomical developments that paved the way to modern investigations.

Modern Astronomy

As early as the mid-1500s astronomers were questioning the concept of an Earth-centered universe, and much current knowledge is rooted in the work of the seventeenth-century astronomers Kepler, Newton, Galileo, and others.

In the year of his death Nicolaus Copernicus (1473–1543) published his final scientific contribution, *De Revolutionibus Orbium Caelestium* (On the Revolutions of the Celestial Spheres). This work stimulated a series of scientific accomplishments that led to an understanding of gravitational forces and eventually to planetary exploration. Unlike his contemporaries, who accepted the Earth-centered Ptolemaic model, Copernicus placed the sun at the center of the solar system. His new model stimulated scholars to question how the sun could dominate the system. Although Copernicus's model, which assumed circular orbits, had obvious deficiencies, it predicted the positions of planets more accurately than previous models.

Twenty years later in Denmark Tycho Brahe (1546–1601), a sixteen-year-old novice playing hooky from academic pursuits, observed a predicted alignment of Jupiter and Saturn. He carefully recorded August 18 as the date when the minimum distance between the planets occurred. He was surprised to learn not only that the prediction based on the old Ptolemaic model was in error by a month but that the predictions of the Copernican model were also off by several days.

Intrigued by the failure of both these models, Brahe undertook to learn more, and in the process became a first-class astronomer. When he secured title to the Danish Isle of Ven, he constructed an observatory at Uraniborg and made frequent positional measurements of the planets. Because telescopes had not yet been developed, Brahe used sextants and quadrants similar to those used at the time for ocean navigation. He observed the planets and stars as points of light and determined the time that a selected object crossed his meridian.

Tycho's efforts were interrupted by the death of his benefactor, King Frederick II, in 1588. Shortly thereafter Brahe was forced to leave Ven. In 1599 Brahe moved to Prague, where he acquired a German assistant, Johannes Kepler, a mathematician twenty-five years younger than he. Before Brahe died in 1601, he named Kepler as his successor. Kepler, who was familiar with Brahe's data, began a concentrated effort to fit the paths of the planets with a self-consistent model. Kepler's persistence and faith in the accuracy of Tycho's observations led to a major breakthrough. The mathematical relationships he developed form the basis that modern-day engineers use to calculate well in advance the path that space probes will follow through the solar system.

Kepler's major contribution was to show that the planets orbit the sun in elliptical paths at predictable velocities. Using the size of the earth's orbit and the length of the earth's year as units of distance and time, Kepler derived a relationship between the period of revolution of the planet and its average distance from the sun. He assumed that the relationship was the same for all planets, and deduced the size of the orbits relative to the size of the orbit of the earth. He also showed that when a planet is close to the sun it speeds up and when it moves away from the sun it slows down, thus providing a clue to the nature of the attracting force. With Kepler's simple model, future locations of the planets could be predicted with much greater accuracy than ever before. No one had obtained any information about the actual distance between any of the planets, however, and despite the advances made by Brahe and himself, Kepler still had no idea of the immensity of the solar system.

One hundred years after Copernicus had been a student in Padua, another resident of the city, Galileo Galilei, studied Jupiter and its moons with a telescope. He observed Jupiter as a disk on January 7, 1610. Galileo's telescope was constructed with small, simple lenses made of poor quality glass. He was nevertheless able to see Jupiter's four largest satellites. The small starlike objects revolved around Jupiter in the equatorial plane at distances 6 to 26 times the planet's radius. Because the periods of revolution ranged from 1.8 to 16.7 days, the shifting positions of the small objects relative to each other and to the center of the visible disk of the planet made it readily apparent that this was a system where small bodies orbited around a larger body. Galileo seized this as significant evidence for justifying the Copernican heliocentric model of the solar system and published his discovery in a small book entitled *Nuncius Sidereus* (The Starry Messenger).

Isaac Newton (1642–1727), born thirty-two years after Galileo's discovery, attended Cambridge University and became familiar with Kepler's work. As a youth Newton became fascinated with predicting the motions of interacting moving objects. As his work progressed, he became convinced that all changes of speed and direction of any moving body were the result of the action of forces. In the case of the planets, he assumed that the motion was due to an attractive force between the sun and each of the planets and attempted to derive the nature of such a force. He found that if he postulated that the force (now called gravity) increased as the product of the masses of the bodies and decreased as the square of the distance between them, he could explain in detail the relationships that Kepler had derived from Brahe's data. In fact, he showed that Kepler's proportionality constant in his relation between period and orbit size actually contained information about the masses of the two attracting bodies.

A general philosophy for explaining natural phenomena was developing at the same time that Newton was formulating the theory of gravitation. A search for cause and effect became the acceptable mode for understanding physical processes. One thing this led to, among others, was an understanding of mass and acceleration within the context of simple everyday tasks. Based on these insights, investigators realized that until some distance separating the planets could be measured directly, no one could determine the mass or velocity of any body within the solar system.

Giovanni Domenico Cassini (1625–1712) was aware of this when he became superintendent of the Observatory of Paris in 1671. He decided to attempt to measure the distance between Mars and Earth. When Earth, in its orbit around the sun, catches up and passes a planet, the alignment is called opposition and the planet appears on the observer's meridian at midnight. At that time Earth is between the planet and the sun, and the distance between it and the planet is at its minimum. Cassini realized that during the coming Martian opposition Mars' elliptical orbit would also place it at a minimum distance from the sun and bring about an exceptionally close approach of Earth to Mars. Because the shapes and relative sizes of the two orbits were known from Kepler's and Newton's work, Cassini argued that measuring the distance between the two planets would enable astronomers to determine the scale of the solar system and thus the actual dimensions within it. He realized that if two individuals located at widely separated latitudes on earth could carefully measure the north-south displacement of the apparent location of Mars relative to nearby stars at midnight, the observations could be combined to determine the apparent shift of the planet among the stars. Then the distance between Earth and Mars could be determined by triangulation.

With this in mind, in 1671 Cassini arranged for John Richer to go to Cayenne, an island at 5° north latitude off the northern coast of South America. Richer's observations were combined with those of Cassini and others at the Observatory of Paris to calculate the distance between Earth and Mars, and from that, to determine the scale of the solar system. Although the measurements had limited accuracy, this was the first time the average distance between the earth and the sun (the Astronomical Unit) had been determined. It was revealed as more than 20,000 times the earth's radius.

The dimensions of the solar system had thus been determined by 1666, and it was possible to convert the angular measurements within the jovian system to actual distances. A reasonable value of the mass of Jupiter could also be obtained from the motion of the Galilean satellites around the planet. Measurements of this sort, based on Newton's formulation of Kepler's laws, indicated that the mass of Jupiter was more than 300 times the mass of the earth and that in comparison to Jupiter the masses of the satellites were very small.

These facts were interesting, but once the distance to Jupiter was known, an even more perplexing property of the planet was revealed—its density. When the apparent angular size of the disk of Jupiter was converted to a linear diameter and the total volume of the planet calculated, the resulting volume was more than a thousand times the volume of the earth and the average density was about one-fourth that of the earth. Logically, the more massive planet should be more compressed than the earth. Our forebears therefore entered the era of modern telescopic observations aware that our giant neighbor could not have the same internal constitution or composition as the earth but must instead be composed of low-density material. By the end of the seventeenth century astronomers had obtained baffling clues about the composition of the solar system and knew that Jupiter was not similar to the earth.

Planetary Astronomy

In the seventeenth century the manufacture of lenses was primitive and the resolution that astronomers could achieve with their instruments was poor. But in the eighteenth century William Herschel (1738–1822), who discovered Uranus in 1781, began to construct larger telescopes with mirrors in place of lenses and to use these telescopes to observe faint objects. As telescope makers became more proficient at grinding mirrors, construction became cheaper and amateur astronomers could afford to buy reflecting telescopes with mirrors that had diameters of 8 to 12 inches.

This diameter seems small by modern standards, but it is adequate for observing details in the cloud deck of Jupiter. Irregular heating and mixing of the earth's atmosphere create eddies that randomly deflect incoming light, creating a rapid jiggling of the various parts of the visible disk. This effect, called seeing, causes stars to "twinkle" and generates an overall smearing of planetary images. Because Jupiter is a bright object and the distorting cells in the earth's atmosphere are usually about 10 inches in diameter, telescopes with larger apertures do not yield greater spatial detail for visual observations.

As the quality of telescopes increased, a growing interest in stellar astronomy and faint objects led many professional astronomers to ignore the planets. As a result, during the late 1800s and early 1900s much of the information concerning seasonal changes in Jupiter's clouds was recorded by amateur astronomers.

During the mid-1900s planetary astronomers consolidated their ideas and carefully considered what they did know about the solar system. A more complete understanding of the nature of the earth and access to improved experimental tools led them to concentrate on gleaning as much as possible from the sketchy information that was available about the nature of the other planets. The knowledge that detailed measurements of the positions of the satellites could reveal how the mass was distributed within the planet encouraged investigators to believe that they could learn more about Jupiter. A growing awareness of how cold this remote planet must be and the fact that a spectrograph could be used to obtain information about its chemical composition led to systematic efforts to determine the nature of the atmosphere. By 1950 Gerard Kuiper and his students at the University of Arizona had determined the abundance of ammonia and methane in Jupiter's freeze-dried environment. Seymour Hess at Florida State University had considered the general atmospheric circulation and compared these extreme conditions to those found on earth. In 1958 Bertrand Peek, the director of the Jupiter Section of the British Astronomical Association, published an observer's handbook that catalogued the variable aspects of the atmosphere. Thus the understanding of the planet had progressed to the point that investigators were more than ready for a closeup view.

Robotic Spacecraft

By 1950 progress in the field of rocketry and technology in general led to a renewed interest in planetary observations. The possibility of obtaining high-resolution data with robotic spacecraft and of eventually sending manned missions and colonizing other planets led to increased interest in and rapid expansion of knowledge of the solar system.

During the 1950s both the Soviet Union and the United States worked to develop earth-orbiting spacecraft, and the launch of Sputnik I (Russian for satellite) on October 4, 1957, marked the beginning of the space age. Explorations of Earth, Mars, and Venus and of the moon were forerunners of missions to the outer solar system. As plans were formulated to explore the outer solar system, Jupiter was the prime target. Two pairs of spacecraft, Pioneer 10 and 11 and Voyager 1 and 2, have now flown by Jupiter, and an orbiting craft, *Galileo,* is scheduled to arrive in late 1995. We will discuss the Pioneer and Voyager missions here and use the interpretations of the data acquired by them throughout the book. In chapter 16 we will review the state of the Galileo Mission.

Pioneer and Voyager represent two different types of spacecraft. Pioneer, as its name implies, is designed to make preliminary explorations. This type of spacecraft is stabilized against tumbling in space by spinning about its own axis. The instruments rotate with the spacecraft and scan the planet as the craft passes it; thus the types of observations that are possible are limited. The Mariner-type Voyager spacecraft is more complex. Its stabilizing system is equipped with three detectors: one points toward the sun, a second to the planet of interest, and the third locks onto a preselected bright star about 90° away from both sun and planet. Once the sensors find their targets, the ship moves through space without rotating. This stability allows for more complex instruments, all of which share a steerable platform and point to the same spot to obtain timed exposures. The Voyager spacecraft costs more, so the two Pioneer craft were sent to Jupiter first. When the Voyager Mission followed, it took advantage of information obtained by the Pioneer.

Pioneer 10 and 11

The Pioneer Mission to Jupiter was officially approved at NASA headquarters in February 1969 and was undertaken by the Ames Research Center at Moffett Field, California. The initial goal of the mission was to fly two spacecraft by Jupiter. As the mission progressed, however, it became apparent that Saturn could also be visited, and planning for the mission was extended to encompass this possibility.

Pioneer 10 was launched on March 2, 1972, by an Atlas-Centaur launch vehicle, and Pioneer 11 followed on April 5, 1973. These spacecraft flew through space with their large dish antenna pointed back toward Earth, spinning at a rate of about five revolutions per minute. The instruments, sweeping across target areas, could be commanded to systematically step perpendicular to the direction of spin, thereby mapping the planet. But the instruments could not stay in one place for a time exposure of any region because they were constrained to a fixed rate of rotation. For this reason no camera was placed on board either Pioneer 10 or 11. Instead the images were obtained with a scanning photometer, an instrument that consisted of a single light-sensitive cell that could be stepped perpendicular to the direction in which the spacecraft was spinning. In this way an image could be scanned line by line. Because the scanning port had to be large enough to allow sufficient light to be collected in the scanning mode, the spatial resolution of the images was limited. It took a long time to acquire each image, and only 23 were transmitted to the earth by Pioneer 10 in the twenty-four hours

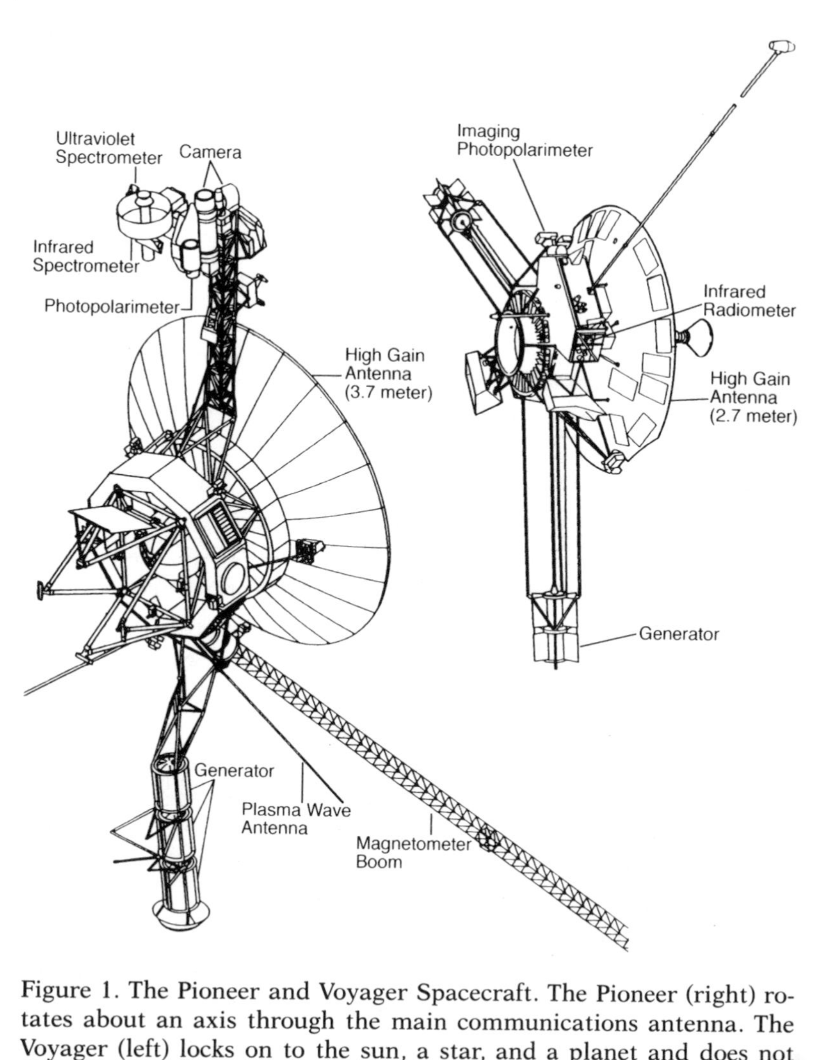

Figure 1. The Pioneer and Voyager Spacecraft. The Pioneer (right) rotates about an axis through the main communications antenna. The Voyager (left) locks on to the sun, a star, and a planet and does not rotate. (Adapted from NASA SP-439 and NASA SP-446)

before its nearest encounter with the planet. Pioneer 11 obtained 17 images during a similar interval. Pioneer images, small in number and with limitations of resolution, revealed little new information about atmospheric motions.

Nevertheless, during these 1973 and 1974 encounters new information concerning the temperature and pressure within the atmosphere was obtained from infrared detectors and from the craft's radio signal. The signal grew weaker as the spacecraft flew behind the planet; then it was lost entirely until the spacecraft emerged on the other side. The manner in which the transmitted signal faded was used to determine how the atmosphere varied with height. The results were consistent with those gathered by the infrared instrument, and as astronomers had suspected, indicated a hydrogen-rich atmosphere similar to that of the sun.

Voyager 1 and 2

During the 1970s the positions of the outer planets in their orbits were such that a spacecraft could travel from one to the next, taking full advantage of gravitational assists to allow cost-effective visits to all five outer planets: Jupiter, Saturn, Uranus, Neptune, and Pluto. Such an Outer Planets Grand Tour would have required several separate craft, with launches in 1976 and 1977 to send ships by Jupiter, Saturn, and Pluto and launches in 1979 to use Jupiter to reach Uranus and Neptune. A scaled-down version of this project finally emerged in 1972 when the Voyager Mission, with two Mariner-type spacecraft, was approved. Originally only Jupiter and Saturn were included in the plans, but later the Voyager team was allowed to extend the Voyager 2 trajectory to include Uranus and Neptune. Pluto could not be attained with this limited effort.

In the pre-Shuttle days of 1977 Titan III-E/Centaur assemblies were used to launch the two Voyager spacecraft. Because Voyager spacecraft were three-axis stabilized, data could be collected more easily with a variety of instruments. Areas of investigation included ultraviolet spectroscopy; infrared spectroscopy and radiometry; imaging science, with two boresighted cameras; and photopolarimetry. Magnetic field and particle detectors were designed to obtain data concerning the interaction of the solar wind with Jupiter's magnetic field and upper atmosphere.

In addition to the advantages of a more sophisticated spacecraft, considerable improvements had been made in the area of communication. The Pioneer Mission was able to transmit in the S-band (with a frequency of 2295 MHz or a wavelength of 13 cm; see appendix 2), while the Voyager could transmit at that frequency as well as in the X-band (8418 MHz or 3.6 cm), which allowed a data rate of 115,000 bits per second. This higher transmission rate resulted as much from improvements in the earth-based Deep Space Network (DSN) facility as in the spacecraft communication system. As recently as 1965 the Mariner 4 Mission to Mars had been limited to the crippling rate of 8.33 bits per second.

The three-axis stabilization accommodated the use of a video camera with timed exposures and the improved telemetry allowed Voyager to send a high-resolution image home via X-band every 48 seconds. In early December 1978, four months before encounter, the Voyager images began to show better spatial resolution than the best earth-based photography, beginning one of the most successful harvests of spatially resolved astronomical information ever carried out.

As the Voyager 1 spacecraft approached the planet, good observations with the other instruments became possible and it was prime time for all the experiments on board. An exploratory effort revealed a faint ring about Jupiter's equator. The small potato-shaped inner satellite Amalthea came into view as the spacecraft approached the planet, but due to the arrival time and the trajectory, high-resolution views of Io, Ganymede, and Callisto were not obtained until after closest approach to the planet. Volcanoes on Io, giant impact patterns on Callisto, and evidence of flooding and stress in Ganymede's crust were evident. Voyager 1 cameras were unable to scrutinize Europa closely, however, and that satellite remained an enigma.

On April 24, 1979, six weeks after the Voyager 1 encounter, Voyager 2 began making observations. This time closest approach to Callisto and Ganymede, as well as the first high-resolution view of Europa, occurred before nearest encounter with the planet on July 9, 1979. This encounter was as successful as the first. Encouraged by the success of the Voyager-Jupiter encounters, NASA continued its efforts to carry out the Galileo Mission to explore the jovian system more completely.

♃ PART II

JUPITER'S ATMOSPHERE AND INTERIOR

♃ Chapter 2

Jupiter's Atmosphere

Through the eyepiece of a telescope Jupiter glows with a pastel pearl-like luster. Against the night sky, the color differences are so subtle that many individuals detect little variation, even when the Great Red Spot is visible. Yet this muted coloration indicates a balance of composition and physical conditions in the visible cloud deck which, together with an impressive assemblage of winds, holds the key to knowledge of the atmosphere. Over the past several centuries astronomers have sought to understand this mystery.

Historical Observations

As early as 1610 observers viewed Jupiter as a luminescent disk with a diameter slightly less than about one-hundredth of a degree. When we consider that Jupiter would appear to be about fifty times smaller than the moon when it is overhead, it is easy to see that reasonably good optics are needed to study the atmospheric details visible on the face of the planet.

The first patterns that early observers discerned were east-west parallel bands extending across the planet. These have been recorded throughout history and are visible in small telescopes. The intensity and color of this banding of Jupiter's atmosphere vary subtly with time and position on the planet. The colors range from pale yellow and brown to bluish and white, and when seen through the earth's unsteady atmosphere, the changes appear like the play of colors in an opal. This contrasts with our current mental picture of Jupiter that has been influenced by high-resolution Voyager images with much sharper spatial definition and so strongly color-enhanced to accentuate the cloud structure that the published colors have little relation to reality (compare plates 1 and 2).

Early observers were also able to track dominant features across the planet, noting, for example, that a specific cloud would cross its face in less than five hours. In 1692, at the Observatory of Paris, Giovanni Cassini reported that equatorial markings moved faster than features at higher latitudes. Observations of this type yielded periods of rotation at low latitudes of 9 hours 50.5 minutes and at mid-latitudes of about 9 hours 55.6 minutes, corresponding to rotation rates of 36.58° and 36.27° per hour, respectively. This difference in rotation rates causes an eastward displacement of equatorial cloud systems of about 7.5° in twenty-four hours relative to clouds located near 20° latitude. If this relative motion is due to clouds drifting in the local winds, the implication is that equatorial winds are blowing eastward at a speed of about 390 km/h (240 mph) relative to the mid-latitudes. Such strong eastward motion near the equator of Jupiter is totally unexpected. Although the earth has prevailing winds, there is no equivalent wind pattern in the equatorial region when it is viewed from a spacecraft.

The Clouds

As early observers became more aware of weather and climate here on earth, they began to wonder about the nature of Jupiter's clouds. They reasoned that because Jupiter is more than five times farther from the sun than the earth and the area over which the light spreads is the entire sphere surrounding the sun, the solar energy per unit area at Jupiter is diluted more than 25 times. Although Jupiter is larger than the earth and presents a larger absorption cross-section to the sun, it also radiates energy from its larger surface area, canceling this advantage. If the absorption of the dilute radiation from the sun and the amount of energy radiated to space is in equilibrium, the average cloud-top temperature must be colder than −150°C (Celsius). If the planet's atmosphere was anywhere near that cold, early observers wondered, what material could form the clouds? Not water: in such a frigid atmosphere, water could not melt and the clouds could not change in the ways observed. Chemists who study ice formation tell us that these low temperatures are so extreme that all the oxygen should have combined with the plentiful hydrogen, forming crystals of water ice that would have sunk well below the visible cloud layer. Chemists predict that the upper level of the planet's atmosphere is completely desiccated, and that the clouds are composed of white ammonia ice crystals, formed when hydrogen and nitrogen combine.

In an effort to understand the chemical composition of the clouds, astronomers began to study the reflected spectrum of the planet. But because it is easier to understand the clouds' chemistry after considering the global circulation of the atmosphere and the nature of the interior of the planet, we will delay this discussion until chapters 5 and 6. Here we will review the winds and their interaction with long-lived cloud systems. In chapter 3 we will look at the Red Spot and similar cloud systems, and in chapter 4, consider the interior of the planet. We will close the section on the planet in chapter 7 by reviewing general models that attempt to explain the winds and long-lived spots.

Cloud Variability

During the first half of the twentieth century an amateur group, the British Astronomical Association (BAA), generated a valuable data bank. Using visual observations, its members collected measurements of the rotation periods and made drawings of Jupiter's cloud structures. Their results were published in the memoirs and journals of the society, and many of the drawings are preserved in libraries of the BAA and the Royal Astronomical Society in London. Along with the publications of the sister organization in the United States, the Association of Lunar and Planetary Observers (ALPO), and reports from the Observatory of Paris, this work formed the historical basis for the photographic observations that followed.

At the same time the BAA and sister societies were recording the variability of Jupiter's clouds, they were developing a system of nomenclature to designate different regions of the atmosphere. They based it on the terminology used for climatic zones on the earth. The broad band extending about 6° on either side of the equator—which tends to be whiter than the nearby regions—was named, reasonably, the Equatorial Zone. This region is bounded on either side by darker regions called the North and South Equatorial Belts. But the nearest regions poleward of these belts, which tend to be white, were given the unlikely names of North and South Tropical Zones. These zones were bounded on the poleward side by another pair of belts and more poleward zones dubbed the North and South Temperate Belts and Zones. This set of belts and zones span about 70° on the planet, extending 35° north and south of the equator. For our purposes, if we are willing to accept terms such as tropical and temperate to describe frozen ammonia ice clouds with temperatures lower than −100°C, this nomenclature is fairly unambiguous. But at higher latitudes the banding is less distinct and such terms as North-North Temperate Belt are cumbersome. Therefore we will not use the BAA nomenclature for higher latitudes, and instead describe them and the polar regions generally or designate a specific latitude.

The Photographic Record

By 1890 photographic emulsions on glass plates were in use, and during the first half of the twentieth century photographic observations were carried out at Lick Observatory, Mount Wilson Observatory, and Lowell Observatory. The planetary archive at Lowell Observatory has copies of much of this early jovian data, including the Lick Collection. It is somewhat disappointing because much of the data is recorded on a yellow-to-green–sensitive emulsion, a range of color within which the cloud deck of Jupiter has the least contrast. Either blue or red reveals more detailed cloud structure. Nevertheless, when these collections of data are compared with more recent records, they reveal long-term changes.

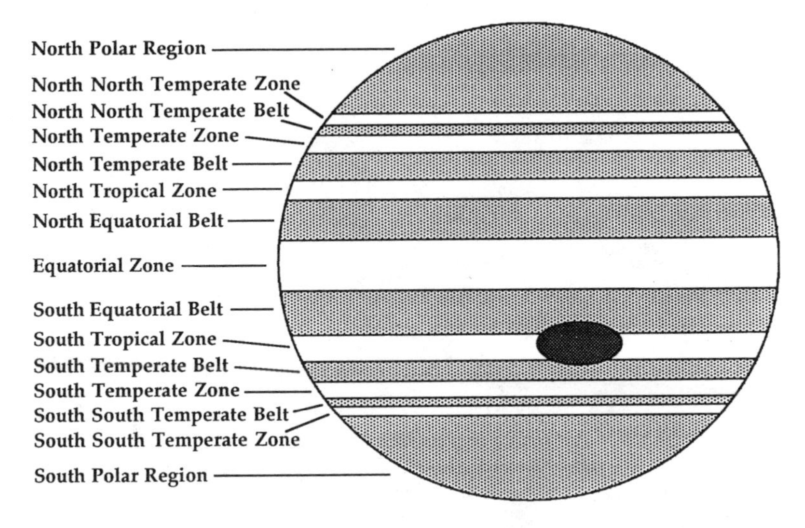

Figure 2. Traditional Nomenclature.

In the recent past observational programs at New Mexico State University Observatory, Lowell Observatory, and Pic du Midi Observatory in France have concentrated on high-resolution color sequences of the cloud deck. By selecting color sensitive photographic emulsions and filters that transmitted broad bands of color, observers could obtain images of Jupiter in near ultraviolet, blue, green, red, and near infrared light with short enough exposure times to capture the planet during intervals when distortion by the earth's atmosphere was at a minimum. These techniques made it possible to obtain a permanent photographic record of large-scale cloud structures in Jupiter's atmosphere. These systematic observing programs contain hundreds of thousands of images that were obtained under controlled, documented conditions. Currently they provide a historical database for a detailed analysis of motions of cloud systems that are larger than 2000 km in diameter.

Electronic cameras are now available that utilize charge-couple devices (CCDs), semiconductor chips with two-dimensional arrays of elements that respond to near-infrared as well as visible light. These cameras are more sensitive than photographic emulsions and allow shorter exposures, making use of

Figure 3. Jupiter over a Span of Twenty-eight Years. From the upper left to lower right, blue-filtered images show Jupiter's variable appearance in 1965, 1968, 1974, 1980, 1986, and 1993. (New Mexico State University)

brief intervals when turbulence in the earth's atmosphere is minimal. Such a capacity enhances the investigator's ability to obtain higher resolution ground-based observations. In addition, the Pioneer and Voyager spacecraft have flown by the planet. Now the total collection of long-term low-resolution images taken from the earth and the short-term high-resolution records from the Voyager spacecraft are being augmented with a long-term series of intermediate-resolution observations with the Hubble Space Telescope. Capabilities of the space telescope and the Galileo orbiting spacecraft, scheduled to arrive in 1995, will be discussed further in chapter 16.

High-resolution observations like the Voyager data set also serve to enhance the historical ground-based observations from which the long-term atmospheric responses can be tracked. An example is the increased understanding of the nature of the winds and of the manner in which the winds influence long-lived storm systems such as the Red Spot.

Figure 4. Six Color Images Obtained on March 8, 1993. In the upper row, from left to right, blue, green, and red filters were used. In the lower row, from left to right, respectively, narrow-band filters are centered in regions of weakly and strongly absorbing methane and near IR. (New Mexico State University)

The Global Circulation

Based on their knowledge of the earth's atmosphere, climatologists are aware that a thorough understanding of global wind patterns is fundamental to understanding the energy balance within any atmosphere. Three fundamental mechanisms can contribute to generating winds in Jupiter's atmosphere: the manner in which the sun's energy is absorbed; the transport of energy from the lower regions by convective mixing or vertical upwelling; and the rate at which heat is lost to space. Because the rate of heat loss is controlled by the ease with which infrared radiation can pass through the atmosphere, infrared (IR) data is needed to help meteorologists interpret visual images. Consequently, astronomers have used the Voyager IR instrument and new IR detectors at ground-based telescopes to measure Jupiter's heat loss.

Jupiter is more strongly illuminated at the equator, therefore the difference in heat absorption from equator to pole should generate a temperature gradient. At the same time the rapid rotation of the planet should interfere with the motion of rising convective bubbles and deflect them into a cylindrical flow around the planet. A combination of these two effects should generate a global circulation with characteristic wind patterns.

Early observers of Jupiter's visible cloud deck obtained measurements of east-west motions of the cloud systems that revealed a complicated variation of the rotation rate with latitude. When the longitudinal positions of other isolated clouds in the jovian atmosphere were compared to the position of the Red Spot, it was apparent that characteristic velocities varied substantially with latitude. Efforts to interpret this complex pattern were hampered by the fact that there was no observable solid surface to provide a reference for measuring the horizontal winds. Without that sort of reference, observers could not understand the general circulation of the atmosphere. Astronomers have found a natural frame of reference in the period of variation in the radio signals, 9 hours 55.5 minutes. These signals are caused by the interaction of the incoming charged solar particles with the planet's magnetic field. This magnetic field should be deeply rooted in the planet; using it as a frame of reference, the circulation of the atmosphere relative to the interior can be studied in much the same way as meteorologists would study the earth's winds relative to its surface. (See appendix 3 for definition of rotation rate and chapter 14 for a description of the magnetic field.)

Ground-based observers were aware that individual clouds appeared to be sheared apart both north and south of certain latitudes, indicating the existence of narrow wind jets at these latitudes. These jets appeared to occur near the edges of the belts and zones, outlining the regions of differing reflectivity. After the longitudinal displacements of all the well-defined clouds in the northern hemisphere were carefully monitored over days and weeks, four eastward jets were identified, located at 6°, 20°, 32°, and 38° planetocentric latitude. (A second system of latitude, called planetographic, has been traditionally used by professional and amateur astronomers, based on the local vertical direction on an ellipsoidal planet. In this system the jets would be at 7°, 22.5°, 35°, and 42° latitudes, respectively. For general clarity, we will use the planetocentric system, which is analogous to the latitudinal system on earth. See appendix 3 for conversion between the systems.)

In the southern hemisphere, an eastward jet near −6.5° and a westward jet at −17.5° were also mapped. However, it was difficult to detect small rapidly moving clouds between −22° and the pole. Although the banded aspect of the two hemispheres was similar, the southern hemisphere contained large well-defined oval cloud systems such as the Red Spot, centered at −20° and the White Ovals, at −30°. Their presence reduced the contrast and masked evidence of any high-speed east-west flow, but their shape and variations in their motions suggested that the wind jets should be there. Thus the Voyager imaging team members planned to collect evidence to solve the problem of the missing wind jets.

Horizontal Winds

Improved measurements of the horizontal wind speeds were obtained from the Voyager images. Because the planet was rotating in front of the cameras, the time interval between images that revealed the same region of the cloud deck was about ten hours, the period of rotation of the planet. Images could be converted to a time-lapse movie, much like those of cloud development in the earth's atmosphere. Image-processing techniques could be used to analyze the archived images and determine the longitude and latitude of every pixel, or data point, on the planet. Therefore it was possible visually to select a set of distinct cloud structures in the images and to determine the longitude and latitude of each cloud feature within each image.

If the dimensions of the planet were included, the difference in longitude of the cloud features in a pair of images separated by a known interval of time could be used to trace an east-west, or zonal, wind. In a similar manner, the difference in latitude could be used to compute the north-south, or meridional, wind. To those of us who studied the Voyager images and carried out this analysis, it was apparent that almost all the cloud motion was in the east-west direction. Using images that spanned the planet from pole to pole, my colleagues and I confirmed that these winds are much stronger than the winds near the earth's surface. Winds in the equatorial region of Jupiter blow from west to east with velocities as fast as 170 m/s (almost 380 mph) relative to the rate that the core is rotating. The maximum equatorial winds are found at 6° to 7° north and south of the equator. The speed of the east wind decreases steadily in the direction of the poles, until at 15° north latitude the clouds are moving westward at speeds of 25 to 70 m/s. North of this latitude the wind speed decreases to zero, then shifts to an eastward direction, reaching another eastward maximum of more than 150 m/s near 20° north latitude. These prevailing zonal winds shear the clouds apart and the strong east-west smearing yields the banded appearance that is seen with earth-based telescopes.

During the early analysis of the Voyager data at the Jet Propulsion Laboratory my colleagues and I became aware of large local variations in the cloud motion that indicated a great deal of turbulence. This turbulence was the result of large cloud masses expanding and rolling in the winds. If all the clouds were sheared apart immediately, there would be little variation in the wind measurements, but this was not the case. When a cloud system forms as the result of strong vertical convection that carries a parcel of gas upward in the atmosphere, the parcel must expand to reach a local pressure balance. This expansion will result in local cooling and formation of fresh white ammonia ice. Analysis of the Voyager data showed that the convective structures tend to form cloud systems that rotate in the wind and translate eastward or westward as a unit until they are torn apart by the local wind shear. The random motion generated by this process confuses the situation, and even though vertical and meridional mixing must play a large role in maintaining the observed cloud structures, these motions are not easily detected in this data. Thus the team was left with a map of the variation of the zonal wind with latitude within a single thin layer of this giant atmosphere and with some information concerning the response of isolated cloud systems to that wind field.

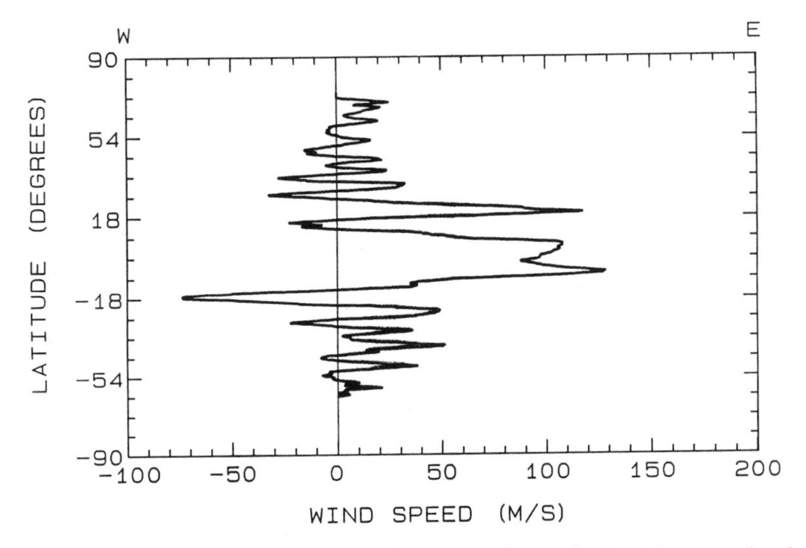

Figure 5. Average East-West Winds Versus Latitude. Positive wind values imply the clouds are rotating eastward faster than the planet's core while negative values represent local westward winds. Differences in speeds and locations of jets in northern and southern hemispheres are apparent.

It is important to remember that the Voyager measurements span only two jovian weeks, and therefore represent a brief snapshot in a jovian season. Even so, the degree of north-south asymmetry in the magnitude of the east and west winds at similar latitudes is surprising. Jupiter is tilted only 3.2° on its rotational axis relative to its path around the sun, and its orbit is only slightly elliptical. With this small latitudinal variation of solar heating with season, it is not obvious that there should ever be such large differences in wind speed in the two hemispheres.

To approach this problem, investigators need to understand how these winds vary with time. My students and I, in collaboration with James Westphal and Ed Danielson from California Institute of Technology, have derived zonal winds from early Hubble Space Telescope images and have found no differences between the Voyager and Hubble measurements. Interestingly, these data were obtained almost twelve years apart and represent the same season in consecutive years. This lack of variation in the winds is necessary but not sufficient evidence to conclude that the magnitude or latitudinal position of the zonal winds does not change. To evaluate the extent of the wind variation, it will be necessary to view the planet in all seasons for several years and to use data from historical ground-based records, the Voyager Mission, the earth-orbiting Hubble telescope, the Jupiter-orbiting Galileo spacecraft, and the anticipated flyby of the Cassini spacecraft on its way to Saturn. Planetary scientists hope that careful analysis of all this information will allow them to obtain a better understanding of how the climate of this gas giant differs from our own.

Although astronomers have obtained no direct information about how energy is transported from the equatorial to polar regions or about wind speeds below the clouds, meteorologists are confident that a careful study of cloud systems such as the Red Spot will provide clues. These large cloud systems rotate in the wind, and their existence requires that the conditions be such that they are not destroyed. In chapter 3 we turn to the oldest known phenomena in Jupiter's atmosphere in our search for clues to understand its dynamics.

Chapter 3

The Red Spot and Other Cloud Systems

The Great Red Spot is centered at 20.5° south planetocentric (23° planetographic) latitude and spans about 23,000 km in the east-west direction and 12,400 km in the north-south direction (the diameter of the earth is 12,750 km). The spot is bounded on the south, or poleward, side by a relatively mild eastward wind and on the northern side by the strongest westward wind on the planet. These driving winds are deflected around the Red Spot, and it rotates about its center in approximately seven days, generating the well-trimmed shape that has fascinated ground-based observers for decades (see plate 2*b*).

Early observers speculated that the Red Spot was locally generated and marked the site of a large obstacle on the buried surface. The British Astronomical Association (BAA) archives contain records of times when the Red Spot was visible. The BAA observers viewed the Red Spot as a whole and were interested in how it moved relative to other clouds. They recorded its apparent size and noted when it was on the central meridian of the planet for many consecutive days each year. They used these data to determine the period of rotation of the Red Spot about the axis of rotation of the planet. I have used the BAA data and my own recent measurements to determine the longitudinal shift of the Red Spot across the face of the planet (see figure 6). From this figure it is clear that the spot was moving faster in 1920–1940 than it does now and that it behaves like a free floating body. Consequently it cannot be associated with any fixed feature on a submerged rigidly rotating surface. The Red Spot drifts at a nearly constant rate for an interval of time spanning twenty to forty years; then, in a matter of months, it accelerates

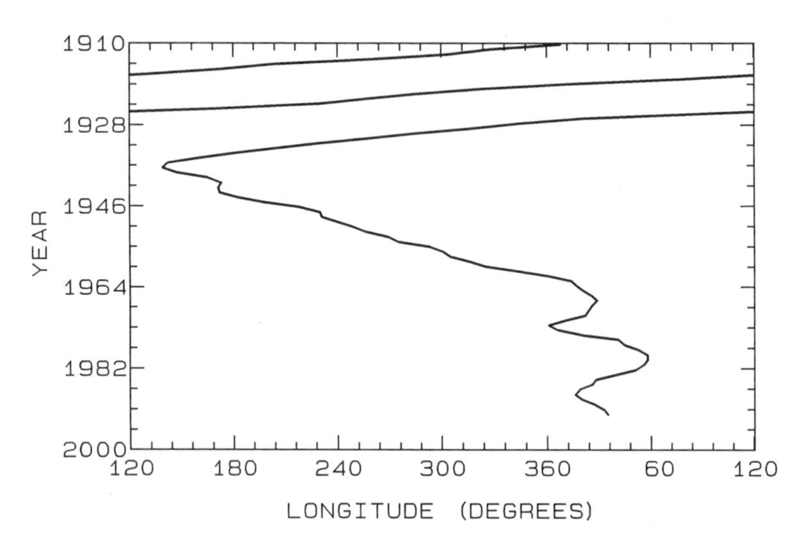

Figure 6. Longitudinal Drift of the Red Spot within a System Rotating at a Constant Rate. Note that in 1910–38 the spot was drifting eastward toward smaller longitudinal values. In the late 1930s the spot slowed down, and since 1960 it has drifted at a variable rate. (The reference system has a rotational period of $9^h55^m40.6^s$.)

or decelerates and settles down to move at another drift rate until it again undergoes a major change in rate. Whatever the mechanism that maintains this feature, such a pattern of motion can only be associated with an atmospheric phenomenon.

These observations do not rule out the possibility that local superheating deep in the atmosphere originally generated a disturbance that pumped heat into the upper atmosphere, where the resulting large-scale turbulence evolved into the Red Spot. At this point observational evidence offers no clue to the origin of the spot but simply indicates that the visible cloud system must have sustained itself without interruption over a period of at least 150 years.

Minor accelerations and decelerations of the Red Spot occur over time spans of a few days to weeks. If a series of measurements of the longitudinal position of the Red Spot are made in the weeks following such an acceleration, a plot of the data will reveal that the spot slows down as if it were suffering a local drag. The spot will tend to return to its former drift rate, indicating that its motion is the result of interaction with the local winds.

When a terrestrial storm forms around a ground-level low-pressure area, the air moves toward the center and generates a central rising column. At the top of the column the material will expand and cool and an equalizing downward flow will tend to develop around the perimeter of the weather system. As the top of this rising column expands, the equatorial side of the high level cloud will have to travel farther to encircle the earth than the part of the cloud that is expanding poleward. Unless the cloud can readily gain energy from its local surroundings, the equatorial part will be moving more slowly than the poleward side and the cloud will tend to rotate about its center. In the southern hemisphere the rotation would be counterclockwise for a rising system.

This comparison to terrestrial storms hints that the Red Spot may have formed when an extra large convective cell, or warm bubble, was carried up from below. If this were the case, however, the analogy provides no assistance in determining the rates of flow, the amount of material being transferred, or the depth to which the feature extends. In addition, the similarity between the Red Spot and terrestrial cloud systems does not rule out the possibility that a comet or asteroid may have collided with Jupiter. The energy of the impact would have heated the atmosphere locally, causing it to expand and interact with the prevailing winds. If the impact occurred near 20° south latitude, the local winds could constrain the disturbance and cause it to evolve into a cloud system like the Red Spot.

A major difference between the Red Spot and terrestrial weather systems is that the storm centers in our atmosphere are not constrained by strong east-west winds but can drift north or south as local conditions vary. These terrestrial storms are driven by evaporation of water at the earth's surface and condensation at higher levels in the atmosphere, and are strongly influenced by placement of oceans and continental land masses. In comparison, the Red Spot exists in a desiccated atmosphere so cold that all water has condensed out well below the visible clouds. The spot is not free to drift in a north-south direction, it is trapped within the prevailing wind field. Photographs in the Lowell Observatory collection illustrate that even though the east-west dimension of the Red Spot has varied by a factor of two during the twentieth century, its north-south dimension has remained nearly constant, extending from 15.4° to 25.4° south latitude.

The recent discovery of the Great Dark Spot in the atmosphere of Neptune reminds Red Spot analysts of another baffling trait of the Red Spot. Voyager movies of the Great Dark Spot reveal that its north-south and east-west dimensions expand and contract, generating a huge "flopping" structure. The Red Spot behaves in a much more constrained manner. In the late 1960s Gordon Solberg and Elmer Reese, at New Mexico State University, reported that the entire Red Spot oscillates back and forth one degree in longitude over an interval of about ninety days. No process capable of provoking such a response has yet been identified. Such baffling behavior is one of many characteristics of Jupiter's cloud features that remains unexplained.

Brightness Variations

Reports of pink or red spots in Jupiter's southern hemisphere date back to the seventeenth century: a drawing by Cassini, displayed at the Observatory of Paris, provides evidence that the Red Spot is more than three hundred years old. But gaps in the eighteenth-century chronological database make it impossible to establish that the spot did not dissipate and then re-form during those intervals. Further, the ability to distinguish the Red Spot from its surroundings varied greatly, leading to confused reports in the historical records. Some sources state that the Red Spot was discovered in 1879, at a time when it was surrounded by white clouds and was very well defined. The observers' excitement led to a sharp increase in published material concerning the nature of the Red Spot, especially during the decade following 1879. But individual drawings and records preserved in the London archives of the Royal Astronomical Society leave little doubt that from 1830 to 1870 the Red Spot was present and looked very much as it did during the Voyager encoun-

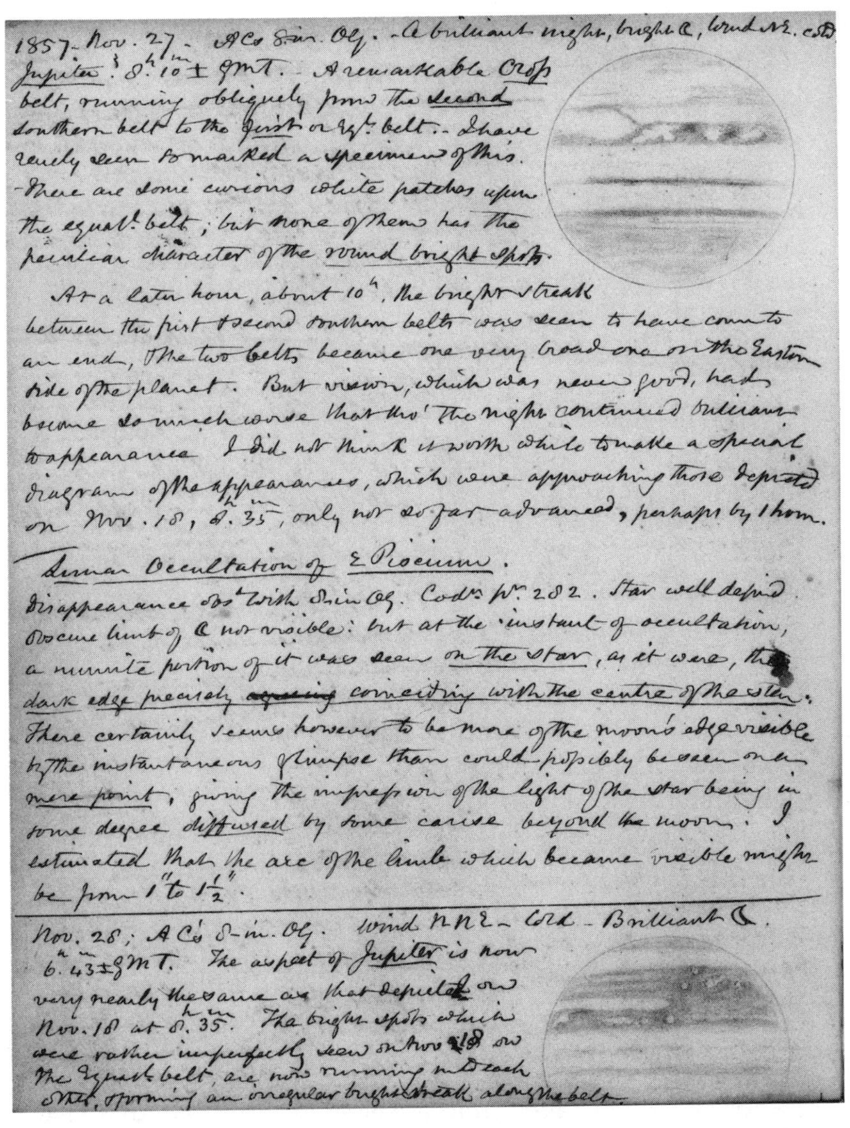

1857 - Nov. 27 - A Co 8 in. Obj. - A brilliant night, bright ☾, wind N.E. cold
Jupiter: $8^h 10^m$ ± GMT. - A remarkable Cross
belt, running obliquely from the second
southern belt to the first or Eqr. belt. - I have
rarely seen so marked a specimen of this.
- There are some curious white patches upon
the equatl. belt; but none of them has the
peculiar character of the round bright spots.

At a later hour, about 10^h, the bright streak
between the first & second southern belts was seen to have come to
an end, & the two belts became one very broad one on the Eastern
side of the planet. But vision, which was never good, had
become so much worse that tho' the night continued brilliant
to appearance I did not think it worth while to make a special
diagram of the appearances, which were approaching those depicted
on Nov. 18, $8^h 35^m$, only not so far advanced, perhaps by 1 hour.

Lunar Occultation of ε Piscium.
Disappearance obsd with 8 in Obj. Codn. No. 282. Star well defined.
Obscure limb of ☾ not visible: but at the instant of occultation,
a minute portion of it was seen on the star, as it were, the
dark edge precisely ~~agreeing~~ coinciding with the centre of the star.
There certainly seemed however to be more of the moon's edge visible
by the instantaneous glimpse than could possibly be seen on a
mere point, giving the impression of the light of the star being in
some degree diffused by some cause beyond the moon. I
estimated that the arc of the limb which became visible might
be from 1" to 1½".

Nov. 28; A Co 8-in. Obj. Wind N.N.E. - Cold - Brilliant ☾.
$6^h 43^m$ ± GMT. The aspect of Jupiter is now
very nearly the same as that depicted on
Nov. 18 at $8^h 35^m$. The bright spots which
were rather imperfectly seen on Nov 18 on
the Equat. belt, are now running into each
other, forming an irregular bright streak along the belt.

Figure 7. Page From the Observing Log of the Rev. W. P. Dawes on Nov. 27, 1857. In this sketch south is at the top, as seen in the telescope. The Red Spot is to the left with white clouds following, much the same as seen in Voyager images. (The Royal Astronomical Society, MS Dawes 3 folio, 154 Verso)

ter. The baffling longevity of this feature has led to a great deal of speculation.

Historically the appearance of the Great Red Spot has varied between two extremes, from being readily apparent because of its color and location to being obscured and white. The Red Spot is most apparent when the adjacent South Equatorial Belt is highly reflective and apparently homogeneous. At these times detection of individual clouds in the westward jet that separates the South Equatorial Belt from the poleward South Tropical Zone is difficult, and the Red Spot is entirely encircled by the highly reflective cloud deck. This contrast is greatest when the spot is observed through a blue filter.

When the South Equatorial Belt is brown and the contrast of the Red Spot is low in blue light, smaller closed eddies with east-west dimensions on the order of 5000 km form in the westward jet at 17.5° S. These eddies move westward relative to the Red Spot and encircle the entire planet in about three months. As they approach the Red Spot from the east, they are swept around its north side. When they arrive at the west end of the spot, the motion of some of the eddies carries them into the flow around the poleward side and into the general circulation around the Red Spot. These eddies carry fresh white ice into the spot, forming a collar of white clouds that is particularly bright in blue light. In 1968 Elmer Reese and Bradford Smith, working for Clyde Tombaugh at New Mexico State University, measured eddies of this kind on photographic plates and verified the counterclockwise rotation of the spot.

During times when the Red Spot is obscured, traditionally called the Red Spot Hollow aspect, the spot will tend to be white and have the same reflectivity as its surroundings. But even at these times the northerly deflection toward the equator is visible as a hollow in the south side of the South Equatorial Belt. The size of this visible feature indicates that the Red Spot does not shrink, but only ceases to be an effective absorber of ultraviolet, violet, and blue light. A comparison of the Pioneer data with Voyager data indicates that the contrast of the Red Spot relative to its surroundings is controlled by the level of turbulence in the westward jet and the rate that the spot is ingesting small white eddies. The Red Spot was darkest in 1878 to 1882, 1927, 1936 to

1937, 1961 to February 1966, and November 1968 to July 1975. In 1989 the South Equatorial Belt became bright for a short time, but returned again to the aspect photographed by Voyager, where the belt is brown and active sites generate bright clouds that expand and sheer apart in the local winds. By late 1992 the belt had brightened again and the Red Spot was well defined and quite dark in blue light; but again, on April 9, 1993, a South Equatorial Belt Disturbance began that led to increased turbulence and a return to the brown, less reflective condition.

The Red Spot Color

The color of the Red Spot is different than that of the equatorial region, another reddish area. The Red Spot absorbs more of the blue, violet, and ultraviolet light than most other features. However, there are small short-lived ovals that occur at similar northern latitudes that share this characteristic. These structures form in a region where only anticyclonic cloud systems like the Red Spot would be stable.

Although small oval features in this northern region display a morphology that is consistent with a rotation where the eastward end is rotating toward the equator (clockwise), the orientation of their main axis always remains tilted in the wind shear and they fail to grow and develop into great northern red spots. Disappointingly, they disappear after two to three years. The imaging system on board the Pioneer spacecrafts recorded a small oval at this latitude. This spot was bright in red light, and like the Red Spot, dark in blue light. Ground-based observations in ultraviolet to near-infrared light verified that the manner in which this little Pioneer red spot absorbed light was very similar to that of the Great Red Spot. Although several oval structures were visible in the zone in the Voyager images, none were as large as the Pioneer spot, nor did they display the same strong ultraviolet absorption. This agrees with historical records that reveal red, brown, and white short-lived oval structures to be characteristic cloud systems within this northern zone.

The fact that some oval-shaped clouds near 18° north and south latitude are darker in blue and ultraviolet light than other clouds has been interpreted as evidence that they contain a trace constituent that absorbs blue light, giving them a reddish appearance. This material could be carried up from the lower atmosphere. At this latitude the vertical mixing could extend deeper, dredging up a particular compound, or the upward motion could be faster, carrying the compound up to visible heights before chemical reactions destroy it. The identity of this absorber is unknown, but my colleagues and I are currently using the ultraviolet spectrographs on board the Hubble Space Telescope in an effort to identify it.

Similar Storm Systems

In 1939 the South Temperate Belt was abnormally bright and contained three brown regions that appeared to be breaks in the encircling band of clouds at 30° south latitude. The intervening white regions extended for tens of degrees of longitude along the belt. As the planet rotated, observers recorded the longitudes of the leading and trailing edges of the darker regions as A–B, C–D, and E–F. The brighter regions attracted little attention until, in 1948, observers realized that the white regions were evolving into well-defined cloud systems. The east-west extent of these three systems continued to decrease and their shapes became more oval.

In the mid-1950s Elmer Reese submitted his observations of the ovals to Bertrand Peek, who was director of the Jupiter Section of the British Astronomical Association (BAA). At that time Reese was still labeling the leading and following edges of the White Ovals as F–A, B–C, and D–E, an artifact from the earlier notation where the limits of the brown regions were A–B, C–D, and E–F. When Peek included Reese's measurements into his database, he confirmed that the regions that had contracted to form the ovals had been consistently observed since 1939, when the first brown breaks in the South Temperate Belt had been reported. The unlikely names FA, BC, and DE were accepted by the BAA and are still used.

As the length, or the east-west dimension, of the White Ovals decreased, the rate of their eastward motion also decreased.

Even so, they are still drifting eastward relative to the Red Spot. Therefore each oval catches up and passes to the south of the Red Spot once in about 2.6 years. Because the velocity of the individual ovals varies, they drift in longitude and interact with each other. In 1975 BC approached FA to within 50° of longitude. When this occurred, the region between them brightened and remained bright during the 1979 Voyager encounters. Following this close approach, FA accelerated relative to BC, and by August 1986 FA was leading BC by 165° and had approached to within 60° of DE. By the fall of 1989 DE had drifted to within 17° of BC, a closer approach distance than ever before; but again BC accelerated and began pulling away from DE.

Voyager images indicated that the three ovals extended for about 9000 km in the east-west direction and spanned about 5000 km, from 27.7° to 32.5° latitude. Like the Great Red Spot, the three ovals rotate about their centers in a counterclockwise direction, rotating once in about five days. The images also revealed turbulent weather systems similar to the structure west of the Red Spot. These cloud systems lie to the northwest of the ovals, and like the clouds west of the Red Spot, rotate in a clockwise direction.

Although the current White Ovals formed in 1939, earlier records reveal that during the preceding years there were two, not

Figure 8. White Ovals FA and BC. A Voyager 2 view of the ovals, separated by 70° longitude, with similar smaller features to the south of the ovals. (JPL, NASA)

three, similar features at this latitude. With the onset of the brightening that formed the current three ovals, the eastward drift rate of the whole latitudinal region increased. Therefore it was not possible to link either of the former storm centers with FA, BC, or DE. However, the earlier storm systems provide evidence that these large long-lived systems are characteristic features of the southern hemisphere of Jupiter.

Internal Motions

Little was known about the circulation within the Red Spot and White Ovals until Voyager 1 approached Jupiter. The motion of small eddies around the perimeter of the Red Spot had established that it was rotating counterclockwise, suggesting that material was rising from the center and expanding outward. In addition, the placement of the oval weather systems with respect to the belts and zones supported the hypothesis that these ovals were regions of upwelling. Measurements of variable longitudinal velocities of the Red Spot and of the ovals had established that they were atmospheric features that responded to their surroundings. Various simplified models were thus created to consider the behavior of closed eddies in a system that included shear in the zonal winds; the predicted behavior for the eddies in the models was then compared with what happens in the Red Spot and White Ovals.

When high-resolution Voyager images were analyzed, the counterclockwise circulation of the Red Spot was confirmed, and that of the White Ovals as well. Rotational winds as high as 100 m/s (almost 200 mph) were measured in the outer portions of both the ovals and the Red Spot. In the ovals, the magnitude of the winds decreased steadily toward the center, but the Red Spot showed a different pattern. There the strong winds were limited to the outer region, or collar, and the motions in the interior, redder part of the spot were small and chaotic, showing small-scale convection, or local mixing. No outward flow was detected inside either the Red Spot or the White Ovals. Measurement of the White Ovals revealed that they were highly similar in size and structure and that, like the Red Spot, all three were swallowing small eddies that approached them from the east. Like large healthy eddies generated in controlled laboratory environments, these great eddies feed on little eddies. Is this how they maintain themselves? Progress toward solving this problem is slow because there are no direct measurements of the physical conditions that exist below the cloud deck and few data concerning how energy is transported upward above these giant eddies.

Some information about the conditions above the Red Spot has been gained from analysis of the infrared data. By selecting portions of the spectrum where the hydrogen or methane molecules absorb the infrared radiation, it is possible to derive temperatures at various altitudes in the upper atmosphere. Michael Flasar and other members of the Voyager infrared team at the Goddard Space Flight Center showed that the gases above the center of the spot were colder than those above the surrounding clouds. These results indicate that the center of the Red Spot is rising to a higher altitude than the surrounding clouds. Because the pressure is lower, the top of the Red Spot expands and supercools. However, much is still not understood.

Dark Clouds

Large brown features are observed near 15° north latitude in the North Equatorial Belt. At the time of the Voyager encounters four brown ovals (traditionally called barges by ground-based observers), were present in the belt. A. Hatzes, D. D. Wenkert, A. P. Ingersoll, and G. E. Danielson, at the California Institute of Technology, determined that the largest barge is rotating in a counterclockwise direction. This motion is consistent with a cloud system rotating in the local winds. If a large cloud system in the earth's northern hemisphere displayed this motion, the storm center would have a central downflow.

Assuming the same is true in Jupiter's atmosphere, the characteristic dark-brown color of these systems can be explained as a deficiency of ammonia ice particles caused by melting as the ices descend to lower, warmer regions. The color of reflected light is determined by the absorption properties of the atmosphere. There is evidence that a brown haze extends over a wide range of altitudes. If the highly reflective ammonia ice particles are removed from the upper regions, most of the light will be absorbed by the haze, forming a dark-brown oval pattern in the surrounding clouds. These holes, or barges, are typical of the North Equatorial Belt, but no equivalent structure has been observed in the southern hemisphere. From time to time ground-based data reveal similar large brown regions in the North Temperate Belt located at 24° north latitude.

The area of the largest brown oval is 10 million km^2 (3.8 million mi^2). The feature not only moves eastward at a rate of 2.5 m/s relative to the planet's core, but its east-west and north-south dimensions also oscillate in a manner such that when the length increases, the width tends to decrease. Thus the visible surface area remains constant as the system drifts eastward. (In contrast, the ninety-day oscillatory motion of the Red Spot occurs in a manner such that the entire cloud system moves as a unit, oscillating back and forth over one degree of longitude.)

Because the four brown barges did not share a common translational velocity, their relative positions changed between the Voyager 1 and 2 encounters. The smallest feature, located east of the others, moved most slowly and was nearly overtaken by the nearest trailing brown barge. Historically, the large barges move faster and overtake smaller ones. When this occurs, the large spot swallows the small one and continues on its way with scarcely a burp. This is typically what happens when a large eddy consumes a small eddy in laboratory studies of turbulence.

At this stage, when investigators consider the asymmetry in the detailed structure of the horizontal wind pattern in the two hemispheres, they ask a chicken-or-egg type of question. Did the presence of the large anticyclonic features in the southern hemisphere modify the local circulation or did the original southern circulation differ from that in the northern hemisphere in such a way that it was conducive to the formation of large anticyclonic storms in the south and large cyclonic systems at lower northern latitudes? It is also possible that the asymmetry in both winds

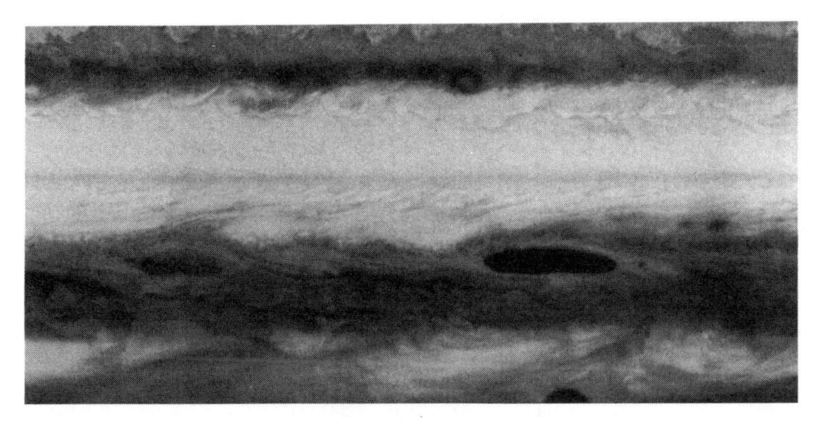

Figure 9. Brown Barge. This Voyager image shows a westward jet located to the north of the barge along the south side of the North Tropical Zone (bright). The wind increases eastward to the south of this jet, generating a strong counterclockwise shear across the belt. (JPL, NASA)

and cloud systems is the result of a more fundamental characteristic of the planet, such as irregularities in the outward flow of the internal heat source. This last possibility, however, is not based on firm observational evidence. Although there are longitudinal variations in the heat flow, no difference in the total heat flow from the northern and southern hemispheres has been observed. If the Red Spot and the ovals are due to a large hot spot that is still in existence, asymmetric heat flow would be expected.

With these phenomena in mind, it is time to turn our attention to the interior of the planet.

♃ Chapter 4

Jupiter's Interior

Knowledge of Jupiter's internal structure and its chemical makeup would greatly enhance the understanding of how the solar system formed. Although there is no way to make direct observations of the interior, indirect clues, such as the 400 km/hr equatorial winds, can provide connections between the planet's interior and its visible cloud layers. These winds cannot be explained by scale models of the atmosphere of the earth, where winds are caused by uneven solar heating of the atmosphere and crust. If Jupiter were like the earth, sunlight would be deposited in the outer layers and the resulting equatorial winds would be pleasant breezes.

New insight was provided in the 1960s when detectors revealed that Jupiter was brighter in the infrared than had been expected. The energy flowing out of the top of the atmosphere was more than one and a half times the amount Jupiter was receiving from the sun. The excess energy had to originate at deeper layers, perhaps even at the planet's core. At last a possible driving mechanism for the winds had been found, but further knowledge of the interior was needed to understand how winds could be generated by this outward transfer of heat.

As is usually the case, finding a partial answer to a problem creates a new question. What is the source of this non-solar energy? The question is interesting in itself, suggesting as it does that by providing its own energy source, Jupiter has some stellar traits. But both question and answer are also part of a set of interlocking pieces of information that can teach us about the unfathomable depths of the planet.

If investigators knew the temperature, pressure, and composition of the interior of the planet, they could compute how much energy would be generated and convince themselves that they understood the planet's interior. Because this is not possible, they must follow another approach. They could try to guess how these quantities vary with depth and then make predictions concerning how much heat their model would lose and whether it would generate the winds that are observed.

Of course, guessing the details of Jupiter's interior from core to clouds would be a frustrating game. A more direct approach can be taken. Careful consideration of all the available observations can be used to understand Jupiter's interior. In this chapter we will consider ways to use these constraining observations to deduce conditions beneath the obscuring cloud layer. The atmospheric "outer boundary," where winds may lead to knowledge about the deeper regions, has already been discussed. But there are other constraints as well, including emission of infrared energy and the structure of the external magnetic field which, as it does for the sun and earth, depends on internal processes. In addition, the degree of flattening of this rapidly rotating giant and the "shape" of its gravitational field can tell us something about how the matter is distributed.

Before scientists can consider these constraints, they must consider how to describe the structure of the interior. Their approach is to develop a computer program that formulates a numerical model that starts at the atmosphere and ends at the core. This is achieved by considering the planet as a set of layers, nested like an onion. At each depth within the planet the temperature, pressure (force per unit area), density (mass per unit volume), composition (ratio of elements and the state in which they are found), and anything else that may help to characterize the nature of the material must be specified.

Observations over the past several decades have provided a wealth of data, both space and ground based, and the speed and storage capacity of computers have progressed at a rate that has impressively outrun inflationary costs, allowing the development of more complex numerical models. As a result, planetary scientists now know a small amount about Jupiter's inner domain. Like the sun, it is composed mostly of hydrogen and helium, and in contrast to the earth, its gaseous region extends deep within the planet. The central temperature is about 20,000°K compared with 7000°K for the earth and 15,000,000°K for the sun. These results derived from the computer-based studies, along with the run of accumulated mass with depth, provide modelers with a framework for specifying the physical conditions within a giant planet and set the stage for understanding the origin of the outer planets. We will consider these points in this chapter and follow with more detail about techniques and a discussion of the insight we have gained through the use of numerical models.

Planetary Formation

Even though a detailed understanding of how heat is generated and transferred to the atmosphere may solve some of the puzzles associated with the jovian winds, there is a larger question to be explored. If scientists understand the interior structure of this planet, they may be able to choose between different scenarios of how the solar system formed.

One possible mode of formation is a straightforward collapse of the solar nebula from a cloud containing essentially solar abundances of all components. If Jupiter formed in this way from material that had the same chemical composition as the sun, then a dense central core would have formed as the heavy elements settled to the center. For a planet of Jupiter's mass (318 earth masses), this core would contain 1 to 1.5 earth masses.

Another scenario for formation of the planets involves two stages of formation. In the first stage, small rocky bodies and icy chunks, composed mainly of water and ammonia ices, coalesce to form an inner core, or protoplanet. In the second stage, the growing protoplanet introduces gravitational instabilities in the surrounding solar nebula that cause a collapse of the local region and the formation of a large hydrogen-helium rich envelope. This scenario would lead to a planet with a dense core of unknown mass that could be larger than that formed from direct condensation of the solar nebula.

Various models of Jupiter that incorporate both these ideas have been considered. These models yield differing predictions for the mass distributions within the interior. If scientists are to choose from among the models and determine the correct chain of events for the formation of the solar system, they need to devise ways to determine the distribution of Jupiter's internal mass. Unlike the earth, where field scientists can install a world-wide set of seismic stations and study the manner in which shocks from earthquakes are transmitted through the interior, investigators of Jupiter's interior are limited to remote sensing from a passing or orbiting spacecraft.

The Gravitational Field

Geophysicists have used spacecraft in near Earth orbit to determine the manner in which mass is distributed below the earth's surface. This is possible because the gravitational force that any unit of mass can exert on a spacecraft depends on both the mass and the distance from the spacecraft. Thus if a dense meteorite were imbedded in the earth's crust, the spacecraft would feel an increased force as it passed over the site. It follows, in general, that analysis of the motions of natural satellites or man-made spacecraft can be used to study the shape and strength of the gravitational field of any planet. The spacecraft must pass close to the planet if this is to be accomplished because at large distances it will sense a gravitational field that appears identical to that generated by a concentrated spherical mass. As a spacecraft approaches Jupiter, an oblate planet, it will feel an extra force tugging it toward the equatorial bulge. As it skims above the clouds, it will also respond to any localized mass concentrations. The total effect of these forces will control the motion of natural satellites or of a spacecraft.

When the motions of the Galilean satellites are used to derive the gravitational field of Jupiter, the field is found to be in close agreement with that predicted by a rotating model with a dense rocky core containing 10 to 20 earth masses. This analysis suggests that if Jupiter formed from a single collapse of a gas cloud of solar composition, then to have a core of this size, the planet must have lost at least 90 percent of the original hydrogen and helium since its formation. Planetary scientists must either propose conditions within the early stages of formation of the solar system that would allow for such a loss or interpret these results to favor the two-stage scenario of planetary formation, with an initial coalescence of a dense core followed by a collapse of enough surrounding gases onto the core to create the outer portions of the planet. Additional clues that would help to distinguish between these scenarios may be found when the current comparison of jovian data with that from the other outer planets has been completed.

Because the natural satellites of Jupiter revolve in nearly circular orbits near the equatorial plane at relatively large distances from the planet, they are not sensitive to mass distributions near the poles or to north-south asymmetries that could be present if there were large internal circulation patterns. Disappointingly, the limitations imposed by the flyby trajectories of Pioneer 10 and 11 and Voyager 1 and 2 spacecraft have added little to the characterization of the gravitational field. Careful analysis of the slight changes in the frequency of the signals transmitted by these spacecraft did not reveal any gravitational anomalies that would indicate large internal convective cells or north-south asymmetries in the field. The data sets from Pioneer and Voyager have been combined to derive refined values of the degree of polar flattening, but hopes for more sensitive data in the near future are not high. The Galileo Mission will provide little additional information concerning the internal mass distribution of Jupiter because its orbit will be near the equatorial plane and outside the orbit of Io.

Future missions that involve spacecraft in elliptical, inclined orbits that pass close to Jupiter for extended durations would provide valuable constraints on the mass distribution within the planet, but no mission of this type is planned for the near future. Thus the best hope is to develop more sophisticated computer models. Let us look at the basic approach in order to understand the nature of this task.

A Numerical Model

If it looks like a horse, feels like a horse, and smells like a horse, you can probably ride it in a parade. This illustrates the philosophy that applies when a computer is used to formulate a numerical model to represent a real thing. The modelers make a reasonable assessment of the problem and apply good working rules to find a self-consistent solution and generate predictions that agree with the nature of the thing being modeled. A numerical model of a planetary interior is based on fundamental assumptions expressed as equations of constraint. The mass, radius, and chemical composition characteristic of the planet are introduced into the model and familiar physical laws are applied. An example of how one of these laws was selected has to do with the fact that the radius of Jupiter has not appeared to vary over the last three hundred years. If the modelers assume that Jupiter is neither shrinking nor expanding, they can incorporate the condition of hydrostatic equilibrium into the model by requiring that the inward pressure imposed by the crushing force of gravity must be balanced by an outward force exerted by the interior's resistance to compression. This would seem to be a reasonable assumption since Jupiter is neither a black hole nor has it blown away since the solar system formed over four billion years ago. The outward expansional forces depend on the temperature and the state of the matter within the planet's interior. Therefore it is easy to see that if the modelers can find enough realistic constraints and access an efficient computer, they should be able to construct a numerical model.

A basic problem in beginning a model is selecting the chemical composition. Spectroscopic observations reveal that the atmosphere of Jupiter is mostly hydrogen and helium, and its low density suggests that this may be the case for most of the planet. Even though the compression within a planet 318 times more massive than the earth must be huge, an average density of less than one-fourth that of the earth indicates that assuming Jupiter is composed largely of light elements, hydrogen and helium, is reasonable.

When equations of constraint and parameters have been defined, a computer program that generates the model can be written in a manner such that the temperatures and pressures are adjusted to satisfy all the equations of constraint at all depths within the model. When this is accomplished, the model is complete and the values of the unknown variables—

temperature, pressure, and density as a function of depth within the planet—can be examined. The model can be tested by calculating the heat flow through the outer regions and comparing it to Jupiter's heat loss. In addition, it is possible to calculate the magnetic and gravitational fields that the model generates and see if they are similar to those observed by Pioneer and Voyager spacecraft.

The construction of a model can be considered an experimental process in which the investigator begins simply and gradually develops more detailed representations of the planet in a step-by-step manner. If the results are in reasonable agreement with observations, the modelers may go back to the computer and vary the conditions within the model slightly to determine which observable quantities are sensitive to the specific assumptions within the model. This will allow them to decide which aspect of the model to improve.

An area where considerable progress has been made recently is referred to as "the equation of state." This equation describes the relation between temperature, pressure, and density. This relation is simple for temperatures and pressures where the material behaves as a gas, as the pressure of a gas is directly proportional to the product of the temperature and density. But a given layer within Jupiter's interior must support all the mass above it; therefore the compressive forces increase rapidly, reaching huge values near the center. The modelers must worry about how to describe the state of the matter and how to handle the computations or deal with this rapid rate of increase of pressure with depth.

Rotational Effects

Up to this point we have ignored the distortions of the planet due to its rapid rotation and assumed that the gravitational force is directed toward the center. But telescopic observations of Jupiter reveal that the equatorial radius of 71,400 km is about 6 percent larger than the polar radius. This observed distortion indicates that rotational forces must be considered. The centrifugal, or distortional, force is directly proportional to the velocity of a particle. This force varies from a maximum value at the equator to zero at the pole because a parcel of gas in the atmosphere near the equator will rotate about the axis of rotation at a rate of more than 45,000 k/h, or 28,000 mph (compared to 1670 k/h for the earth), while the rotational velocity of a particle circling the planet near the pole will be nearly zero. Centrifugal force tends to throw the material outward, perpendicular to the axis about which any small volume within the planet is constrained to rotate. This force combines with the gravitational force in a manner that, combined with the internal thermal pressure, allows the planet to expand into an ellipsoidal shape.

Any model of the interior that includes the effect of rotation must predict a response to the centrifugal force that is similar to Jupiter's. Because of the manner with which the centrifugal and gravitational forces vary throughout the planet, adjustment of the mass distribution within the planet will cause different distortions resulting in greater or lesser oblateness. The centrifugal force increases in proportion to the distance from the axis of rotation and does not depend on the internal mass distribution. In comparison, the gravitational force is proportional to the total mass that is distributed interior to that location within the planet and decreases radially outward. Therefore, for a planet with a given mass, volume, and rate of rotation, the more the mass is concentrated into a small dense core, the smaller will be the rotational distortion. A good model will have a mass distribution such that there is enough material far enough from the center to result in a 6 percent distortion. With this discussion in mind let us take a brief look at the historical development of this form of study.

Interior Models

The earliest models of Jupiter's interior were developed by R. Wildt, at Yale University, in the late 1930s. This early work, which utilized simple models and was constrained by limited computing capability, addressed the problems of determining the central temperature and pressure of Jupiter. During the

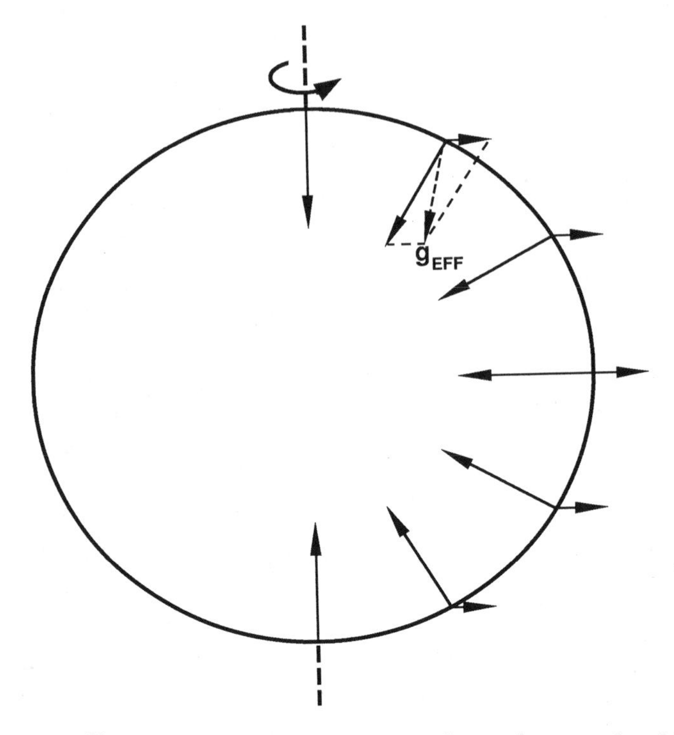

Figure 10. Effective Gravity on a Rotating Sphere. The centrifugal force increases from zero at the pole to a maximum at the equator and combines with the gravitational force to create an effective gravity that is not directed toward the center of the planet. The ratio of gravitational to centrifugal force is about 10 at the equator and increases to 20 at 60° latitude.

1960s and 1970s more refined models were produced by William B. Hubbard, at the University of Arizona, and P. J. E. Peebles and Roman Smoluchowski, at Princeton. One of the questions that arose during these investigations concerned the extent to which the rapid rotation of the planet would tend to divert heated cells of material that were carrying energy outward into a cylindrical flow, symmetric about the axis of rotation. The significance of this is that if such a flow were to form, it could generate the observed equatorial winds and the planet's banded appearance. In addition, if this motion extended down into the conductive region, it would generate a dynamo action that would produce a magnetic field in a manner similar to that in the sun.

Unfortunately, detailed modeling that will lead to a general understanding of Jupiter's magnetic field is in its infancy; huge efforts will be required to develop the detailed computational procedures to represent these interactions. Many of the physical parameters, such as electrical conductivity of the material in Jupiter's core, are so little known that it is difficult to generate a reliable model. This does not mean that progress will not be made, only that it will be hard won and slow.

In order to fit Jupiter's gravitational field, models must incorporate a central dense core that is 10 to 20 times more massive than that of the earth. The remaining mass is hydrogen-helium rich. The models require a central temperature of 20,000°K (three times higher than that of the earth) to match the observed rate of heat loss from the planet and to balance the gravitational force.

The fact that Jupiter is losing a large amount of heat relative to its solar supply requires that the planet have a compensating energy source. One possibility is primordial heat that has not been lost through the 4.6 billion years of the solar system's existence. The initial heating would have resulted from the conversion of the collisional energy to heat as smaller bodies coalesced to form the planet. Just how much heat was generated depends on the balance between accretion and rate of cooling to space, and the maximum temperature reached during the early phases is not known.

Even though primordial heat seems to be a plausible energy source, investigators should look for other processes that could have the same effect. If they assume that the dense core is partially composed of meteoritic material, which seems reasonable, then a possible heat source could be the decay of naturally occurring unstable isotopes of heavy atoms. The use of laboratory data and calculations of the expected yield from isotopic decay, however, predicts a heat loss that is about one-tenth of 1 percent of Jupiter's actual heat loss.

There is yet another possible energy source, conversion of gravitational potential energy to heat. Even slight deviations from pressure balance would allow contraction of the planet.

This would release energy through the conversion of gravitational to kinetic energy. If all the energy lost through heat flow is generated from potential energy released as the planet contracts, the rate of contraction required to compensate for the current energy loss is only a few centimeters per year. Thus the decrease of radius that would have occurred over the last three hundred years could not have been observed by ground-based astronomers. This result does not rule out primordial heat as a possible source of the observed infrared heat flow, but it does indicate that models should be formulated to vary with time or to evolve. These models must include physical relationships that provide for transport of energy from the interior to the surface and compensate for the loss of energy from the interior.

Planetary scientists' understanding of the energy transport yields the following model. In Jupiter's outer atmosphere the density is so low that little energy can be transferred convectively and direct radiation of heat and light plays a large role in determining the low cloud temperatures. This cold atmosphere radiates primarily in the infrared, and the rate of cooling at any level in the atmosphere is controlled by how transparent the region is to infrared radiation. These processes dominate down to a level where the local gas pressure is about 1 bar (a region with pressures similar to those near the surface of the earth).

Deeper in the envelope that surrounds Jupiter's interior the material behaves as a gas, and temperature gradients are relatively large. Convection, or turbulent mixing, transports the heat efficiently. The more efficient this transport, the easier it is to formulate a reasonable, simple equation of constraint. The question of how energy is transported in the deep interior, however, is not as readily answered. Theoretical studies of the behavior of hydrogen indicate that at the temperatures and pressures of deeper levels, hydrogen ceases to behave as a gas and convection is sharply reduced. About a quarter of the way in toward the center of Jupiter, at pressures of 3 million bars, the degree of ionization and compression of the hydrogen becomes great enough that its properties resemble those of a metallic liquid. Still deeper within the rocky core, conduction processes would control the rate of cooling.

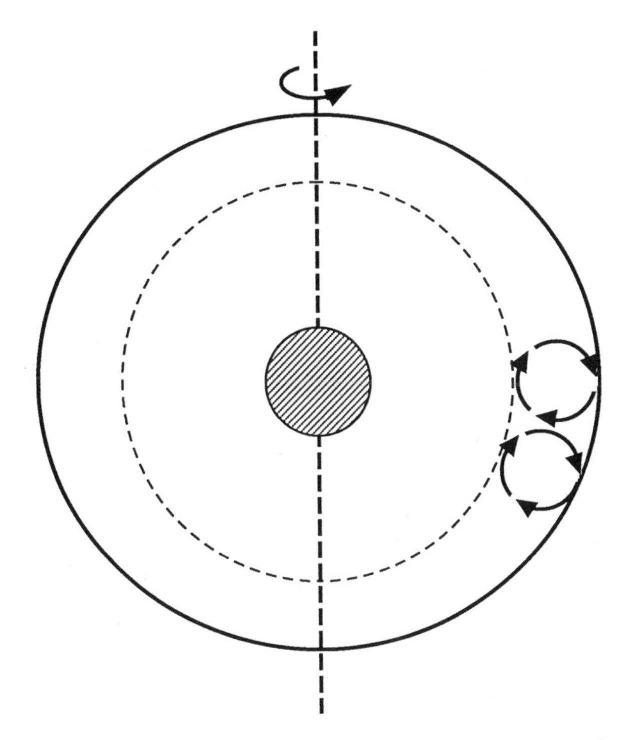

Figure 11. Schematic of the Interior of Jupiter. The central shaded region represents a dense core of 10 to 20 earth masses (pressure = 100 million atmospheres, temperature = 20,000°K). In the region between the core and 0.78 radii the pressure ranges from 40 to 3 million atmospheres. Convection will dominate in the outer envelope and rising gases could be deflected into a cylindrical flow by the rapid rotation.

The best models indicate that Jupiter's central pressure is about 100 million bars and that the planet has a rocky core, surrounded by a convective envelope composed mainly of molecular hydrogen.

Although investigators are not expecting a wealth of new observational data, there is still a great deal that can be done to improve the general understanding of the structure of Jupiter's interior and the manner in which energy is transported into the outer cloud layers. Laboratory studies and theoretical considerations will improve knowledge of the characteristics of the compressed material in the deep interior. With improved numerical models running on high-speed computers, detailed formulations of energy transport and the manner in which magnetic fields are generated can be considered. These investigations will allow us to obtain a more self-consistent picture of the interior of the planet and to understand more about how internal activity manifests itself in the visible cloud layer.

2⅃ Chapter 5

Temperature and Cloud Structure Versus Height

In chapters 3 and 4 it became apparent that ground-based and Voyager images reveal very little direct information about the conditions above or below Jupiter's visible cloud layer. Fortunately there are other ways to learn more about the upper regions of the atmosphere. The manner in which the light from a background star or the radio signal from a spacecraft is modulated by the atmosphere as the star or craft disappears behind the planet and emerges on the other side offers a clue concerning how quickly the atmosphere thins above the clouds. Careful analysis of these data yields information about how the temperature changes with height. In addition, analysis of infrared spectra provide information about the composition and conditions of this outer region. These two independent ways of probing the atmosphere give planetary scientists confidence that they have at least a tenuous knowledge of Jupiter's upper atmosphere.

In this chapter we will consider how ground-based astronomers obtained information about atmospheric temperatures when Jupiter occulted, or passed in front of, stars and how, armed with that data, they used laboratory information and numerical models to estimate where clouds of differing composition would form. We will also consider the results obtained when the Pioneer and Voyager spacecraft disappeared behind the planet. In chapter 6 we will consider the results from spectroscopic studies.

The way that temperature varies with height in a planetary atmosphere will influence the pattern of circulation and is, therefore, basic information that meteorologists must have to understand the weather. If atmospheric scientists have temperature distribution for Jupiter's atmosphere, they can construct models that allow them to determine how the solar energy is deposited in the atmosphere. This understanding lets them compare Jupiter's world with that of the earth and other planets.

Atmospheric Probing

Before the Pioneer and Voyager flights astronomers were aware that if they could measure the light from a star as Jupiter moved in front of it, these data could be used to determine the degree to which the starlight was refracted, or deflected, as it passed through Jupiter's atmosphere. They knew that the extent of refraction at any height depends on the density, and if they had the density as a function of height, they could assume that the atmosphere was in pressure balance and derive the pressure and temperature distributions. To do this, they needed to know in advance when Jupiter's motion around the sun would carry it in front of a bright star. When such an event was predicted, several groups of astronomers traveled to places where this occultation could be observed. Using local or portable equipment, they attempted to measure the star's light as it was occulted by Jupiter. Acquisition of these data was fraught with uncertainties such as bad weather and unstable sources of power to drive the photometers. Despite such obstacles, however, analysis of the observations revealed that the temperature of the visible clouds was about 140°K (−130°C) and that the temperature increased inward and decreased steadily with altitude above the visible clouds (see appendix 4 for temperature scale). At still higher altitudes, in a manner similar to that on the earth, the temperatures increased again, forming a warmer layer at the top of the atmosphere.

Chemical Equilibrium

These observational efforts were complemented by attempts to develop computer models that would aid in determining the composition of the cloud deck and the overlying atmosphere.

Investigators used the observed temperature distribution, extending the inwardly increasing trend to below the visible cloud level. They assumed that the atmosphere was in pressure balance and that all the atoms and molecules in the gas phase were in thermal equilibrium with each other. Then they set up a scheme that allowed them to use laboratory data to estimate which molecules and ices would be present.

Calculations of this kind, carried out by John Lewis and Ronald Prinn, at the Massachusetts Institute of Technology, and by others, predicted a multilayered cloud system and indicated that the visible clouds were ammonia ice, forming at a depth in the atmosphere where the temperature was about 140°K (−130°C). Above these clouds the hydrogen-helium atmosphere contained small amounts of methane and ammonia gas and traces of other molecules composed of hydrogen, carbon, and nitrogen. Below the ammonia layer two other regions of condensation were predicted. At a depth where the pressure is slightly higher than at sea level in the earth's atmosphere (1 bar), the models predict that hydrogen sulfide (H_2S) and ammonia (NH_3) react to form ammonium hydrosulfide (NH_4SH) clouds. Still deeper, at a pressure of nearly 5 bars, clouds of water ice should form.

These models assumed that the chemical composition of Jupiter was the same as the atmosphere of the sun, and the calculations included the interaction of sunlight with the clouds and the overlying gases. Below the cloud deck the atmosphere was assumed to be actively convective. This was consistent with the observed variability of the cloud structures.

These earlier studies provided a basis for interpretation of the visible cloud deck even though their findings did not always match with or explain ground-based observations. Ammonia, ammonium hydrosulfide, and water form white ices, for example; therefore the models offered little immediate insight into the nature of the coloration of Jupiter's clouds. Later spectroscopic observations turned up several conflicts with the predictions of the models: hydrogen sulfide was not observed and water vapor was much less abundant than predicted. In addition, observations revealed that phosphine (PH_3), carbon monoxide (CO), and hydrocarbons such as acetylene (C_2H_2) and ethylene (C_2H_4) were more abundant than the models indicated. Nevertheless these models did provide a basis for understanding the local jovian environment and for designing spacecraft experiments.

Color Variations

When attempting to understand Jupiter's observed coloration, it is important to remember that the variation of colors and reflectivity is due to the interaction of sunlight with the upper atmosphere and the underlying clouds. A region containing both molecular and atomic hydrogen, atomic helium, and small fractions of ammonia and methane with a sprinkling of small particles, or hazes, lies above the clouds. Each of these constituents interacts with the solar radiation in an individual manner. Molecules of hydrogen do not absorb visible light and they interact very weakly with infrared light; however, hydrogen is so abundant above the cloud deck that it still plays a significant role in the infrared. Because helium does not absorb visible or near-infrared light, its presence is also difficult to determine directly. But laboratory studies indicate that it does collide with hydrogen, causing hydrogen to absorb a broader band of infrared colors. This allows spectroscopists to detect helium indirectly. Less abundant hydrogen-rich molecules such as ammonia and methane absorb in the infrared, red, and ultraviolet, creating systematic, identifiable patterns in the spectrum.

In addition to absorbing specific colors of light in the jovian atmosphere, molecules are strong scatterers of ultraviolet and blue light, scattering the light out of the line of sight while allowing a greater portion of the long wavelength red light to be transmitted. If hazes, or aerosols, are present, they will scatter the light differently than molecules. Small particles may exhibit strong color dependence, preferentially interacting with bluer light. These particles do not scatter radiation as symmetrically as molecules, but tend to scatter strongly forward along the direction that the incident light is propagating. Aerosols, or hazes, of these particles may form high in the atmosphere as a result of photochemical reactions that occur when molecules interact with ultraviolet light. For example, ammonia is dissociated when it absorbs an incident ultraviolet photon. This reaction produces amidogen radicals (NH_2), which react with each other to form hydrazine (N_2H_4), which condenses in the atmosphere. Condensation of hydrazyl (N_2H_3) also occurs. Similar reactions convert methane (CH_4) to acetylene (C_2H_2), ethylene (C_2H_4), and more complex hydrocarbons. At Jupiter's low atmospheric temperatures these complex molecules form solids. The solid particles will slowly settle; as they sink, they will encounter other particles, coalesce, and grow, creating an extensive photochemical smog above the cloud deck. Light reflected from naturally occurring smogs of this sort tends to be brown and yellow. Various concentrations of these absorbers and the presence or absence of white ammonia ices along the line of sight into an atmosphere will produce a wide range of tawny yellow-brown shades (see plate 1*a*).

By the early 1970s investigators were convinced that they had a rough idea of the chemistry of Jupiter's atmosphere. The range of yellows, oranges, and browns that the clouds displayed could be loosely explained. However, problems remained with a few cloud structures at mid-latitudes, such as the Red Spot, where ultraviolet and violet light is strongly absorbed. As noted in chapter 3, a larger rate of upwelling or a greater depth of circulation could carry molecules up from below faster than they can be destroyed locally. If this were the case, then phosphine, sulfur, and other ultraviolet absorbing molecules become likely candidates for explaining the short wavelength darkening of the Red Spot. So far, however, none has been shown to be the culprit.

Radio Probing

Before more detailed models could be constructed, atmospheric scientists needed a better understanding of the temperature as a function of pressure (or height) and of the detailed structure of the opaque cloud layers. With this goal in mind, several groups designed experiments that would fly on board the Pioneer and Voyager spacecraft.

When a spacecraft disappears behind a planet, the signal from the craft must pass through the atmosphere on its way back to earth. As the craft moves behind the planet, the homeward-bound signal encounters denser and denser regions in the atmosphere and undergoes more and more refraction. When a signal enters an interface between regions of different density at an angle, the extent to which the beam is deflected is proportional to the difference in the local density. Because the lower layers of an atmosphere must support the upper layers, the force of gravity causes the pressure to increase rapidly (exponentially) with depth. The density of the gas will increase in the same manner; therefore, as a spacecraft moves behind a planet, the rays that penetrate to greater and greater depths are increasingly refracted. The result is that the signals follow curved paths through Jupiter's atmosphere and the signal from the craft is still detectable after the spacecraft has disappeared behind the planet.

The desire to take maximum advantage of this effect contributed to the trajectories that were selected for the Pioneer and Voyager flybys. These preplanned paths were such that from the point of view of the earth-based receivers, each spacecraft would disappear behind the planet and reappear on the other side. Careful monitoring of the modification of the homebound radio signal would allow the engineers to measure the refractivity of the gases as their line of sight moved down through the atmosphere.

This experiment was carried out on all four Pioneer and Voyager spacecraft. But the conditions of the Voyager experiment were more advantageous because the spacecraft transmitted at a higher power and at two frequencies, the S-band at a wavelength of 13 cm, and the X-band at 3.6 cm. The stronger signal allowed the radio scientists to track the spacecraft to greater depths in the atmosphere, and the fact that refractivity is different at different frequencies gave them a more reliable solution.

During the Voyager 1 occultation, on March 5, 1979, the spacecraft disappeared behind the planet at 12° south latitude and emerged in the equatorial region. The Voyager 2 occultation occurred on July 10, 1979, entering at 67° and emerging at 50° south latitude. This near-grazing path provided information about the upper atmosphere closer to the poles.

If the chemical composition is known, the extent of the refraction depends on how the temperature varies with altitude. G. F. Lindal, from the Jet Propulsion Laboratory, and his co-workers considered the fact that the density would be proportional to the

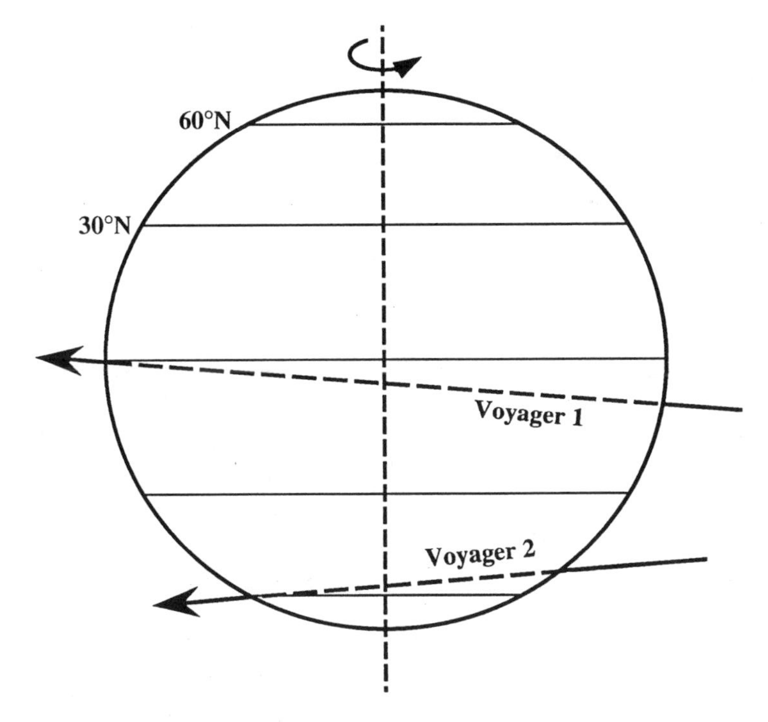

Figure 12. Geometry of the Voyager 1 and Voyager 2 Occultations. The dotted lines indicate the projection of the spacecraft's path perpendicular to the observer's line of sight. Voyager 1 passed within 6 to 8 jovian radii (R_J) and Voyager 2 from 18 to 20 R_J.

local temperature and pressure and also included the atmospheric oblateness due to the rotation of the planet. They assumed that at a given latitude there were no horizontal variations in the atmosphere and the ratio of hydrogen to helium was the same as in the sun's atmosphere, 89 percent hydrogen and 11 percent helium. When these investigators compared the times of arrival and strength of the signal to what would be predicted for a solid airless planet of the same size, they were able to determine the extent to which the radio signal had been refracted. From this they deduced the density as a function of altitude. Then, assuming no vertical motion in the atmosphere, they started at the top of the atmosphere and computed the force per unit area, or pressure, that the gravitational field would exert on

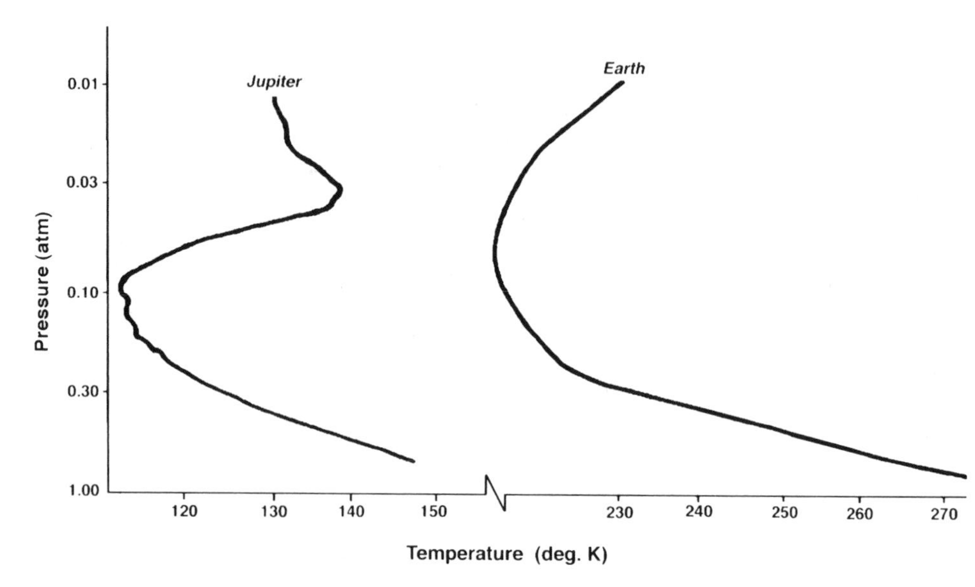

Figure 13. Temperature as a Function of Height in the Atmospheres of Jupiter and Earth. Jupiter's temperature profile was derived from the Voyager 1 radio occultation at 12° south latitude. A typical profile for the earth has been included for comparison.

all the material above a given point. Because this material is a gas and the local pressure is directly proportional to the product of the temperature and the density, the investigators could obtain the temperature as a function of altitude in the atmosphere.

This analysis revealed that the temperature decreases with height from about 150°K (−120°C) near the cloud level to 115°K (−160°C) at about 40 km above the clouds. Above 40 km the temperature increases again. Because the upper layers of the atmosphere should lose energy to space, it seems logical that the temperatures below 40 km must be determined by the manner in which heat from the interior is transported outward through the atmosphere. But above 40 km, absorption of sunlight at ultraviolet, visible, and infrared wavelengths by methane molecules and aerosols, or smog, must cause local heating and determine the temperature structure. Thus these results allow us to consider the dual roles of an internal heat source and solar heating of the upper levels.

Although this cold oxygen-free atmosphere seems alien, conditions within the earth's atmosphere are comparable. The transparency of the earth's atmosphere to visible light allows a major portion of the sunlight to penetrate to the surface, where it is absorbed and heats the atmosphere from below, creating the troposphere, a region where temperature decreases with height. At the same time our atmosphere absorbs ultraviolet and infrared radiation, modifying the temperature at higher altitudes within the stratosphere. Thus comparison of these data from Jupiter with those from the earth moves us a little closer to understanding the widely different family of atmospheres within the solar system.

♃ Chapter 6

Spectroscopic Analysis of the Atmosphere

The science of spectroscopy began in 1814, when Joseph von Fraunhofer became aware that a series of dark lines were visible in the spectrum when sunlight was passed through a narrow slit and dispersed with a prism. Because a spectrum consists of a series of multicolored overlapping images of the slit, an excess or deficiency of light at a specific color, or wavelength, results in a brighter or darker image of the slit or a "bright or dark line" at that color in the spectrum. Comparison of spectra of known elements obtained in the laboratory with those of the sun revealed that the cooler layers of the atmosphere of the sun contained neutral and ionized atoms and small molecules of various elements and that they selectively absorbed and scattered light with specific wavelengths, or colors, from the radiation that was flowing out from lower, hotter regions of the solar atmosphere. By comparing the positions and relative strengths of the lines in laboratory spectra with those in the solar spectrum, the atoms and molecules that were absorbing the light as it passed through the cooler atmospheric regions were identified and the composition of the solar atmosphere was determined. Spectroscopists thus became aware that they could obtain information about the composition of an atmosphere from a remote vantage point.

The sunlight that reaches the top of Jupiter's atmosphere enters and is transmitted through that atmosphere until it is absorbed or reflected. Therefore, by obtaining spectra of the sunlight that Jupiter has reflected and comparing it with a solar spectrum, spectroscopists can determine precisely which colors or wavelengths are absorbed by Jupiter's upper atmosphere. The pattern of the absorbed light indicates the composition of the planetary atmosphere. As early as 1863 Lewis Rutherford reported absorption features in the red end of Jupiter's visible spectrum at wavelengths of 6190 and 6450 angstroms (the average human eye is sensitive to wavelengths ranging from deep violet at 3500 Å to red wavelengths at 7500 Å; see appendix 2 for definitions of wavelength and angstrom). These red absorption features were neither present in the solar spectrum nor readily identifiable in the laboratory.

The lines in the spectrum of an atom are formed when electrons are stimulated to jump from one allowed energy level to another. Molecules are composed of two or more atoms that share electrons, forming atomic bonds between them. Although the electrons within a molecule can become excited in a manner similar to those in atoms, molecules have other more efficient absorbing modes. The energy carried by the photons of light can stimulate the molecules to rotate about their centers-of-mass and to vibrate along the axes of their atomic bonds. The energy required to cause the molecule to vibrate is about ten times greater than that needed to induce rotation. Therefore, when a given photon of light has sufficient energy to stimulate the molecular vibration, photons of slightly bluer light, carrying slightly more energy, will stimulate both the vibration and a series of associated rotations. The resulting observable effect is that each molecule absorbs unique wedges of radiation out of the spectrum and produces a characteristic pattern at specific colors that can be used to identify the molecule that is absorbing the sunlight. The molecules that caused Rutherford's absorption features were not identified until 1932, when Rupert Wildt showed that these features and patterns that had been observed in the infrared were absorptions due to molecules of ammonia and methane. The identification of ammonia with three hydrogen atoms and methane with four hydrogen atoms strengthened the arguments that Jupiter's atmosphere was hydrogen rich.

The discovery that ammonia and methane exist in Jupiter's atmosphere, combined with the knowledge that these molecules absorb even more strongly in the infrared, led ground-based astronomers to develop spectrometers equipped with infrared sensitive photoelectric detectors, forerunners of the infrared instrumentation on board the Voyager spacecraft. For this mission

Rudolf Hanel and his team at the Goddard Space Flight Center designed a dual instrument that consisted of a radiometer and a Michelson interferometer, a type of spectrometer that is described in appendix 5. Both instruments were fed by a compact telescope with a 50-cm (20-in) highly reflective gold-coated primary mirror. A small secondary mirror, mounted in front of the primary, focused the light through a hole in the primary, and mirrors directed the light to the two instruments. The radiometer responded to the visible and infrared light that was reflected from the planet and was used to determine the magnitude of Jupiter's internal heat source and to map local temperature variations on the planet. Its companion, the interferometer, was designed to measure the chemical composition and temperature structure of the atmosphere above the cloud deck.

Heat Loss

As the planet rotated, the infrared telescope scanned swaths north-to-south across the planet, mapping it in infrared light. The sunlit side was mapped on the inbound journey, and the spacecraft was programmed to look back and scan the dark side of the planet after nearest encounter. When the two data sets from the radiometer were compared, there was very little difference in the local cloud temperatures on the daytime and nighttime sides. This is not surprising because Jupiter is cold and far from the sun. The rate of heat loss is a strong function of temperature (see appendix 6 for a detailed explanation), and during Jupiter's five-hour days and nights there is little chance of drastic heating or cooling.

By comparing the intensity of infrared light that was obtained during daylight and dark intervals, Hanel and his team could separate scattered infrared sunlight from infrared radiation that was emerging from below the clouds. They showed that Jupiter emitted energy at a rate of 0.0033 W/cm^2. This means that Jupiter is radiating 1.67 times as much energy as the atmosphere absorbs from the incoming sunlight or that 0.67/1.67, or 40 percent, of the total energy that Jupiter loses must come from its interior.

Spectrographic Results

The final form of the data from the infrared interferometer is a spectrum that reveals how observed light varies as a function of color. Infrared spectroscopists express the color, or related energy, of light in "wave numbers," not wavelength as astronomers have traditionally done for visible light. The wave number is directly related to the energy of the photons and is the number of waves over a unit distance; thus, the bluer the light, the higher the wave number. The Voyager interferometer operated at wavelengths 7 to 100 times longer that of green light (or 3000 and 200 wave numbers, which corresponds to wavelengths of 3.3 to 50 microns—see appendix 2).

The interferometer was designed so that the aperture spanned a 0.25° field of view. Although this allowed far better resolution of individual cloud features than is possible with the largest ground-based infrared telescopes, the resolution was low when compared with the narrow-angle camera. The infrared instrument could obtain only four non-overlapping samples within the field of view of the narrow-angle camera. Because of sensitivity limitations, the observations with the interferometer were begun only three weeks before encounter. At closest approach the most spatial detail was obtained when the field of view of the spectrograph spanned about one-hundredth of the diameter of the planet.

Before the arrival of the Voyager spacecraft the infrared science team did a detailed study of the anticipated appearance of a spectrum of Jupiter's atmosphere. Using solar abundances of the elements and the best laboratory data available for the wavelength and strength of absorption due to the various molecules, they developed a computer program and predicted the spectrum within the wavelength range of their instrument. This study used the temperature versus height relationship that had been derived from ground-based and Pioneer occultation measurements. When Voyager 1 arrived at the planet, the spectra that were returned were similar to what had been expected; the team set to work to analyze the data and determine the chemical abundance and variation of temperature over the disk.

When using high-resolution spectra to probe to various depths

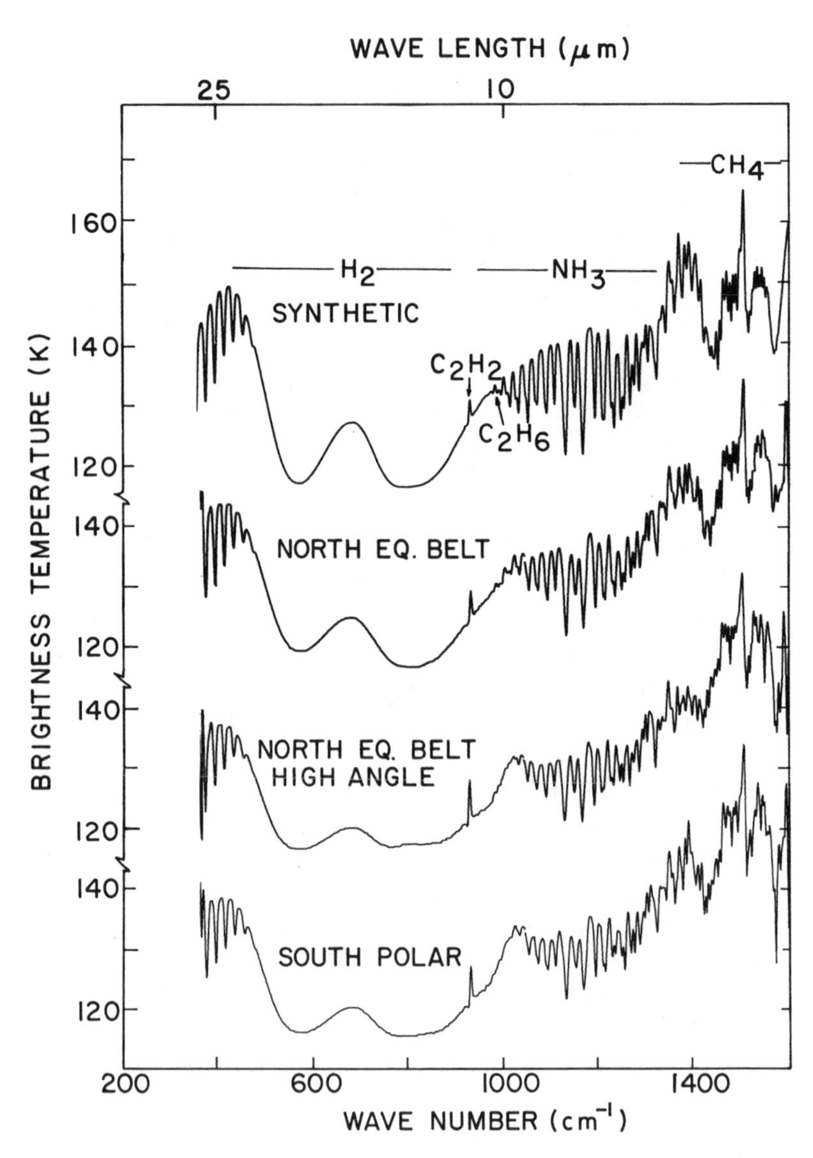

Figure 14. The Long Wavelength Infrared Spectrum. Voyager infrared spectra are compared with a synthetic spectrum generated from a model atmosphere. Spectral features due to absorption by atmospheric constituents are identified. (Courtesy R. Hanel and *Science,* Vol. 204 [1979], 973; copyright 1979 by American Association for the Advancement of Science)

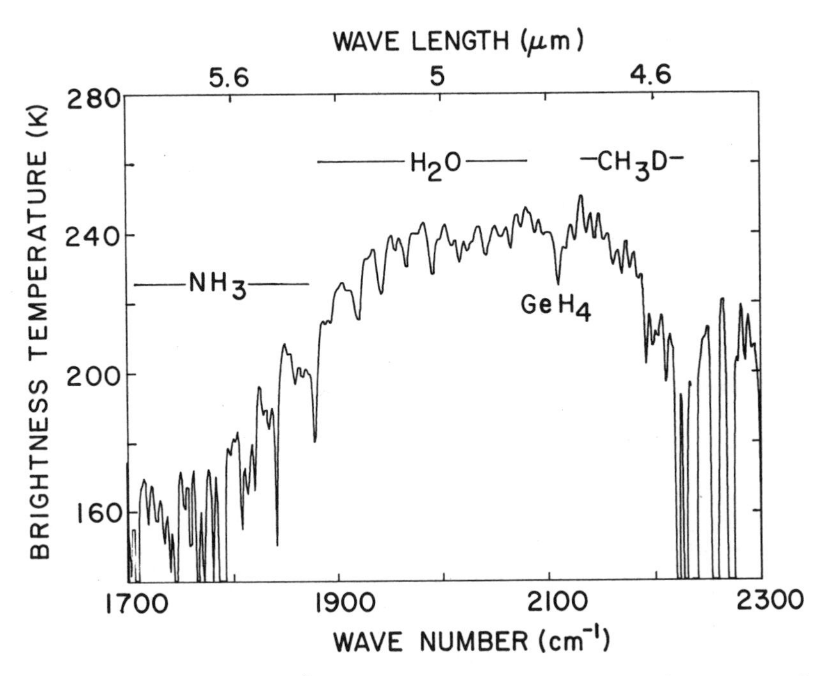

Figure 15. A Shorter Wavelength Spectrum of the North Equatorial Belt. This five-micron region can be used to search for water vapor absorption in the deeper regions from which the radiation is emerging. (Courtesy R. Hanel and *Science,* Vol. 204 [1979], 973; copyright 1979 by American Association for the Advancement of Science)

in the atmosphere, the basic idea is that the viewer is seeing photons that are emerging from different levels in the atmosphere. These heights vary as a function of color, or wavelength, depending on how severely the atmosphere absorbs light at a given color. When the line that is formed by absorption at a given wavelength is very dark, the absorption is strong. Photons of this color of light that are emerging from greater depths in the atmosphere have little chance of avoiding absorption and surviving to emerge from the top of the atmosphere. It follows, then, that the photons that are emerging at wavelengths that are dark in the observed spectrum have come from near the top of the atmosphere. On the other hand, those wavelength regions that are bright in the spectrum have less absorption and photons are emerging from deeper in the atmosphere. A careful analysis reveals that because the density, or number of absorbers, increases rapidly (exponentially) with depth in the atmosphere, at a given wavelength the emergent light originates from a very limited range in altitude. The lowest level from which a photon is likely to emerge is controlled by the increase in absorbers that can destroy the photon. The upper limit is determined by the fact that the exponential decrease of density with height drastically reduces the number of absorbers that can interact with the radiation. Determination of the region of the atmosphere that is being probed is made by careful analysis of the absorption properties of the molecules. This information is obtained from detailed laboratory work and computer modeling of the problem. Rudolf Hanel and his co-workers determined the variation of temperature with altitude in the atmosphere for different regions on the planet. These temperature profiles compared favorably with the radio occultation results, showing the temperature decreasing with height in the troposphere and an overlying warmer stratosphere. In addition, variations of the temperature profiles from place to place on the planet revealed large enough differences to help atmospheric scientists evaluate the forces that drive the winds.

Molecular Hydrogen

In the infrared region (between 12.5 and 33 microns) there are two large dips in the spectrum. These features are due to molecular hydrogen absorption. Although the electrons in a molecule of hydrogen are very symmetrically arranged, causing it to have little response to the variations of the electric and magnetic fields that are associated with the incoming light, collisions with atoms of helium and other molecules temporarily distort it, allowing the hydrogen molecule to more readily absorb the incident light. Because these absorption features in the spectrum are caused by collisions, their strength and width are complex functions of the local conditions. More than 99 percent of the colliding particles in an atmosphere with a chemical composition that is the same as that of the sun would be hydrogen molecules or helium atoms. The team used a variable mixture of

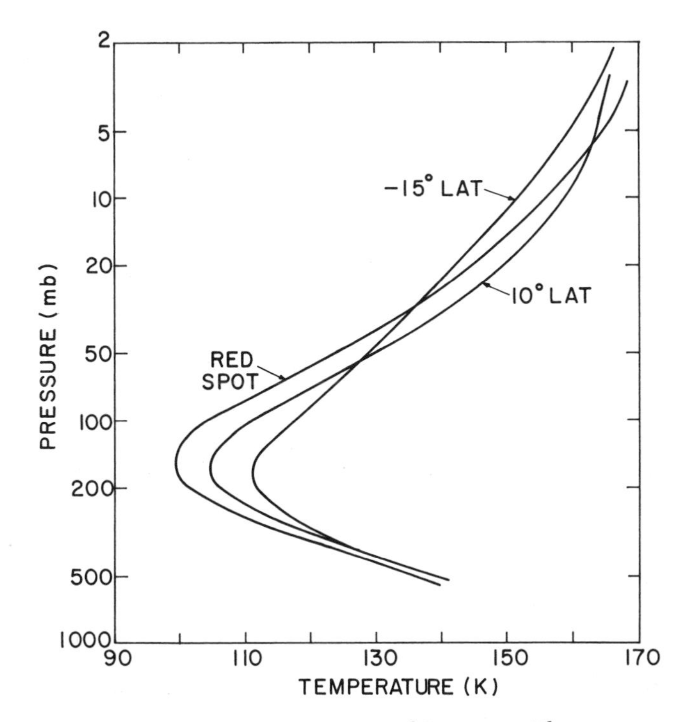

Figure 16. Temperature as a Function of Pressure. These temperature profiles have been derived from spectra in the North and South Equatorial Belts and above the Red Spot. They are similar to those derived from the radio occultation data. (Courtesy R. Hanel and *Science*, Vol. 204 [1979], 973; copyright 1979 by American Association for the Advancement of Science)

these two constituents to derive the model that best fits the data. This model required 11 percent of the colliding particles to be helium and the rest, hydrogen molecules.

Features in the spectrum match laboratory spectra of methane, acetylene, ethane, ammonia, phosphine, water, and germane, an unexpected molecule containing the rare heavy atom germanium. Analysis of the spectra indicates that the atmosphere becomes opaque in the infrared at about 600 millibars of pressure, about 0.6 times the pressure at sea level on the earth. This opaqueness is due to condensation of ammonia and increasing opacity due to the effects of collisions of hydrogen molecules.

The condensation of ammonia leads to a reduced amount of ammonia gas in the atmosphere above the cloud deck. The temperatures in the atmosphere of Jupiter are too high, however, for condensation of methane to occur; thus, absorption features of methane gas with varying strengths can be used to probe the upper atmosphere. Analysis indicated that above the cloud deck methane made up about 0.25 percent of the molecules and ammonia contributed about 0.02 percent. Acetylene and ethane are produced by chemical reactions in the upper atmosphere that are induced by absorption of ultraviolet light. In contrast, water, phosphine (PH_3) and germane (GeH_4), can be transported from deeper regions of the atmosphere. In order to understand the dynamics of the upper atmosphere, analysis is continuing that involves detailed modeling efforts that combine the imaging and infrared data.

Observation of Helium

Although the infrared data provided direct confirmation that a large amount of hydrogen was present in the atmosphere of Jupiter, the only evidence for helium was its role in the way hydrogen absorbed infrared light. Helium, with two electrons so tightly bound that high-energy collisions or energetic photons are required to excite them, is inactive in this cold environment. Ionized hydrogen and helium atoms in active regions on the sun, however, produce excess ultraviolet emission. When this radiation enters the cold upper atmosphere of Jupiter, hydrogen molecules break up. The resulting hydrogen atoms and resident helium atoms absorb the ultraviolet energy and become excited. The probability that they will remain excited is very small. As they drop back to their unexcited level, they emit the excess energy and scatter it in all directions. This process, called resonance scattering, allows direct observations of both hydrogen and helium. Even though this is an ongoing process, the resulting emission from Jupiter's atmosphere cannot be observed from the earth because of the opaqueness of the earth's atmosphere at ultraviolet wavelengths shorter than 3800 Å. Although the Pioneer spectrometer was used and the earth-orbiting International Ultraviolet Explorer (IUE) has been functional for many years to study the ultraviolet region of Jupiter's spectrum, the Voyager spectrometer was capable of isolating smaller regions on the planet and of detecting radiation with shorter wavelengths where hydrogen and helium have strong emissions.

Because the amount of ultraviolet light emitted from solar flares or active regions will vary greatly with time, investigators must be able to measure this variation along with Jupiter's emission. With this in mind Lyle Broadfoot, from the University of Arizona, and his Voyager team designed a combination system with one detector that could measure the variability of the emitted sunlight while a second system measured emission from Jupiter's atmosphere. The spectrometer was designed to reach far enough into the "hard UV" to observe hydrogen (near 1216 Å) and helium (near 584 Å) and to determine their relative abundances.

As Voyager 1 approached the planet, the expected hydrogen and helium emissions were detected. Unexpectedly, other emission lines (at 685 and 833 Å) were also detected. Laboratory comparisons revealed that these lines were from highly ionized atoms of sulfur and oxygen. These were atoms that had been ejected from Io. Trapped in Jupiter's magnetic field, they interacted with the incoming solar ultraviolet light or collided with neutral atoms and emitted their own characteristic signature.

Along with the unexpected sulfur and oxygen emissions, the observations indicated that the hydrogen and helium emissions were quite strong near the poles, where an aurora, or northern light, was visible. Aurorae are generated when high-speed ions of hydrogen and helium from the sun enter the atmosphere near the magnetic north pole and collide with local atoms and molecules, exciting them and causing them to emit light. This auroral light complicates the problem of deriving the helium abundance, but it is possible to separate the auroral and resonance emission by obtaining data on both the day and night sides of the planet. The auroral component obtained on the dark side can be removed from the bright side data. When this was done and the helium-to-hydrogen ratio was determined, the resulting value confirmed that the jovian ratio is quite similar to that for the sun.

Future Outlook

Planetary scientists' ability to obtain additional high-resolution data in the near future is hampered by the fact that the initial instrument package on board the Hubble Space Telescope does not contain infrared instruments (the earliest possible installation of an IR instrument is 1997). The Galileo spacecraft carries a radiometer and spectrometer. The Near Infrared Mapping Spectrometer (NIMS) will resolve smaller cloud features than can be isolated in the Voyager data. But in addition to the severe data-rate constraint, the infrared instruments are not designed to carry out observations with high-spectral resolution at longer wavelengths where the hydrogen molecule absorbs infrared light. The Voyager data revealed that relative strength of the two large absorption features varied with latitude. This variation was interpreted as evidence of vertical circulation. Does the vertical circulation also vary with time? It is unfortunate that the *Galileo* infrared instruments had been designed before the Voyager investigation was completed and the significance of these observations was not apparent. Even though atmospheric scientists look forward to obtaining near infrared observations in the next decade, new advances in the longer wavelength region will have to be postponed.

♃ Chapter 7

Ongoing Atmospheric Research

Although Voyager images of the turbulent clouds to the west of the Red Spot serve as a favorite illustration for articles and textbooks on chaos, in truth atmospheric scientists do not know enough about the fundamental processes that supply and distribute the energy throughout Jupiter's atmosphere to make good use of chaos theory to attack specific atmospheric problems. Much of the current understanding is based on incomplete data sets. Only the first expedition into this realm has been completed.

Even though the spectacular clouds associated with the Red Spot and its documented longevity fascinate us, there are still very basic aspects of this frigid atmosphere that are not understood. Among these are the high-speed eastward winds near the equator, which have no counterpart within the atmosphere of the earth. Historically the region encompassed by these winds, extending seven degrees on either side of the equator, has tended to appear bright and white, but like the earth's atmosphere, Jupiter's appears to display aperiodic variability. There have been two periods of more than a jovian year (11.86 years) when the white clouds disappeared and the region appeared orange and darker than normal. From 1869 to 1883, and again from 1960 to the early 1980 post-Voyager era (6.5 jovian years later), the zone was abnormally orange.

At the time of the Pioneer and Voyager encounters the equatorial zone shaded from white in the north to orange in the south. The Voyager images revealed cloud displacements that corresponded to eastward winds of 110 m/s (1 m/s corresponds to 2.2 mph) along the northern edge, 90 m/s near the equator, and a surprising 170 m/s along the southern edge of the zone. Independent cloud systems form in the northern and southern parts of the zone. Large white plumelike patterns with convective sites near their eastern extremity were present in the north, while small, strongly sheared orange clouds with chevronlike structures were aligned along the southern edge of the zone. Why this difference? Does the presence of the Red Spot modify the southern half of the zone? These questions remain unanswered.

There are many other questions with no current answers. Why does the South Equatorial Belt, located just south of the Equatorial Zone, change from brown to white when the North Equatorial Belt, at similar northern latitudes, does not? Why are dark-brown cyclonic storms like the brown barges never seen in the southern hemisphere? Why are anticyclonic systems like the Red Spot, White Ovals, and smaller oval storm systems located at 40° latitude common in the south, while the only similar storms observed in the northern hemisphere occur north of 45° latitude? Why are the magnitudes of the zonal winds different in the northern and southern hemispheres (see plates 1–3)?

All the above questions remain unanswered. Although the Hubble Space Telescope and the Galileo probe and the instruments on the orbiter will provide new data, this is not the only route to tackling these problems. Scientists can make use of basic natural laws and the increasing capabilities of high-speed computers. Before leaving Jupiter to survey its family of satellites and the surrounding environment, we will consider three areas of study where investigators have or are attempting to shed light on the questions above.

Zonal Wind Asymmetry

No explanation for the large degree of north-south asymmetry in the cloud morphology and wind fields has yet been articulated. The mechanisms that generate these asymmetries are currently unknown. It is not even known whether these differences originate deep within the atmosphere or whether they are generated at shallow depths; investigations are continuing, however.

At a given latitude, Jupiter's clouds experience little variation in the intensity of sunlight throughout a jovian year. In contrast, the 23.5° tilt of the earth's axis is the major cause of seasonal temperature variations. The earth orbits the sun, spinning like a giant top, with its axis of rotation always pointed toward the north star. Because of this orientation, the sun's rays strike the earth at a greater slant angle in winter than in summer, spreading the sunlight over a larger area. This effect, combined with the fact that in winter the tilted aspect of the axis leads to fewer hours of sunlight, generates a large temperature variation in alternate hemispheres as the earth proceeds around the sun. Jupiter's axis is tilted only 3.1°; therefore little variation in solar heating and minimal seasonal variation occur.

A large seasonal effect can also be generated if a planet is orbiting the sun in a highly elliptical or elongated orbit. Jupiter is not: the total range in the distance between Jupiter and the sun is only ±5 percent of the average value. It thus appears that seasonal effects cannot play a role in generating the observed north-south asymmetries. A closer look, however, reveals that the effect of the axial tilt and the eccentricity of Jupiter's orbit combine in such a way that the planet is closest to the sun when the days are longest in the northern hemisphere and farthest from the sun when the days there are shortest. This means that the northern hemisphere will experience maximum summer temperatures and minimum winter temperatures. In the southern hemisphere the short days occur when the sun is at a minimum distance and the long days coincide with maximum solar distance. These two small effects tend to cancel each other out, so that seasonal variation in the southern hemisphere is negligible.

Can the different effects of the planet's axial tilt and eccentricity of orbit cause the observed north-south asymmetry? Certainly experience with the earth's weather would lead meteorologists to expect that storms would be strongest in Jupiter's northern hemisphere. But here the large storms are in the hemisphere with least seasonal variation. Maybe these cloud systems are not analogs of terrestrial storms. Storms on earth are generated by global temperature differences. What if jovian storm systems are produced when an exceptionally large convective bubble is carried up from below? Then maybe the reason they are long-lived in the southern hemisphere is the lack of seasonal effects, whereas in the north, although small, the variation in atmospheric conditions is enough to destroy them.

If this is so, and the large southern storms are driven by enhanced heat flow, observations should reveal excess infrared emission in the southern hemisphere. However, neither Pioneer nor Voyager infrared detectors measured any difference in the energy that was being emitted from the two hemispheres. How can this be explained?

Andrew Ingersoll, who was on the Pioneer infrared team, and his graduate student Carolyn Porco, now at the University of Arizona, addressed this problem. They considered that Jupiter could adjust to seasonal variation in solar heating of the upper clouds in a manner that was different from Earth. Their research indicated that at a given latitude the amount of heat flowing outward from the deep interior could compensate for cooling or warming of the upper cloud layers.

Ingersoll's and Porco's research explains how different latitudes can receive and absorb different amounts of sunlight and not radiate different amounts of infrared energy. It predicts that local convection does not alter the infrared output. When a localized series of bubbles of gas rises from a warmer spot below the visible cloud deck, the strong zonal winds tend to spread the excess heat around the planet, warming the upper region and decreasing the outward flow of internal heat. This damping mechanism modulates any effect that active convection has on the net infrared emission. Although this model accounts for the observed lack of variation in the infrared, it explains neither the marked north-south asymmetry of the zonal (east-west) wind pattern nor the presence of large anticyclonic storm systems in the southern hemisphere. Further, it discourages additional infrared observations as an avenue toward the solution of these questions. There are other ways to gain insight into these problems, such as models that describe how the convective energy is carried up from below and deposited in the atmosphere.

Outward Heat Transport

The Pioneer flybys in 1973 and 1974 reminded everyone that the understanding of the giant planet was scanty at best. Turbulent cloud patterns near the poles and the infrared observations suggested that if detailed studies were carried out concerning the manner in which heat is transported, new insights into how the heat flow from the interior affects the air circulation of this planet might be gained.

In 1976 F. H. Bussé, from the University of California at Los Angeles, demonstrated that rapidly rotating convective spheres can develop unique circulation patterns. He used a device composed of a solid conducting sphere mounted on a rotating axis. This small sphere was encased in a larger crystal sphere so that he could fill the intervening volume with fluid. When he heated the inner sphere, a radially dominated circulation was set up in the fluid, and when the device was rotated, the flow was deflected about the axis of rotation. Bussé's laboratory experiments revealed another interesting phenomenon. Along with a concentric flow, the fluid tended to form a series of cylindrical columns that rotated about their own axes and were aligned parallel to the axis of rotation of the sphere. This resulted in an eastward flow along the outer side of the columns and a westward flow along the inner side. Bussé suggested that a similar flow might occur in the deep convective atmosphere of Jupiter and that this flow might be the source of the zonal winds.

This model has attractive characteristics. For instance, this sort of energy and momentum transfer can generate a large eastward equatorial wind. On the other hand, it is not clear that it would apply to Jupiter's interior, where the density must increase rapidly with depth.

To test whether Bussé's theory holds for deep within the planet, it is necessary to utilize a computer and generate numerical models. Gary Glatzmaier, at the Los Alamos National Laboratories, has written computer programs that attempt to do this. His early work has shown that his numerical models do indeed have strong equatorial winds and a series of alternating westward and eastward wind jets poleward of the equatorial jet. These models represent the deep atmosphere and the interior of the planet down to its dense core. Glatzmaier is trying to describe the winds below the visible cloud layers; he is also calculating the magnetic field that the internal motions of his

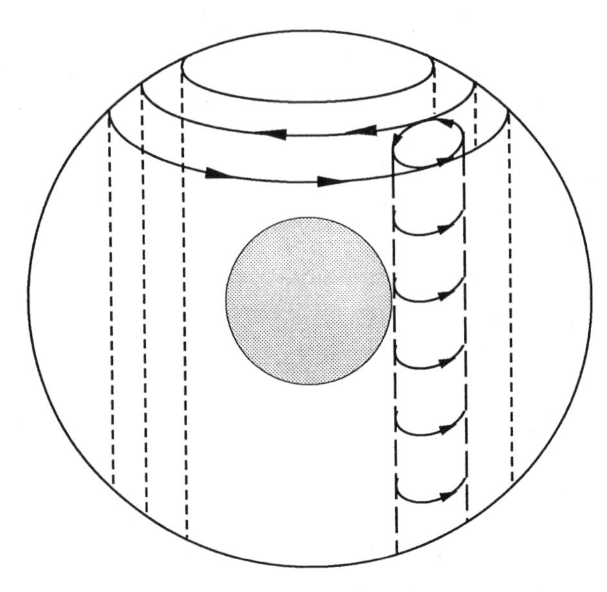

Figure 17. Bussé's Model. Laboratory experiments and computer models indicate that rapid rotation with convection can develop cylindrical flow patterns in Jupiter's interior. This model could provide the mechanism for driving the zonal winds.

model will generate to see if it is as strong as that of Jupiter (see chapter 13).

If Jupiter's atmospheric winds are driven by deep convection, the zonal winds should be insensitive to changes in the outer visible layers of the atmosphere. This would mean that Jupiter's meteorology is quite different from the earth's and that the cloud systems that are observed are trapped within the internally driven zonal flows. Models such as Bussé's and Glatzmaier's—internally driven and highly idealized—predict complete north-south symmetry in the zonal wind pattern, but the fact that the speed of the zonal winds at specific latitudes differs in the northern and southern hemispheres indicates that the true picture is more complicated (see figure 5).

The actual case may be a combination of strong constraints imposed by the deep organized convection and modulation by larger than average aperiodic bursts of energy that emerge from the deeper, warmer regions. How often do these disturbances occur—every 20 years or every 500 years? Who knows? Is the observed meteorology driven by deep winds that vary little with time? Is the interior circulation dominated by cylindrical flow? In the case of the Red Spot, is there slow dissipation of a burst of energy that made its way up through the clouds hundreds of years ago? Was the formation of the White Ovals in the 1930s just a small secondary "burp" associated with the "big one" that generated the Red Spot? As investigators seek ways to answer these questions, they wonder whether the long-lived storms can be studied and understood in a way that will provide insight into the physical conditions deep within the atmosphere.

Modeling the Red Spot

Andrew Ingersoll, at the California Institute of Technology, and Gareth Williams, at Princeton, carried out studies to understand the stability of large eddies in Jupiter's atmosphere. They showed that large oval eddies can be maintained in regions where the zonal winds decrease toward the pole of the planet. One of Ingersoll's students, Timothy Dowling, now at the Massachusetts Institute of Technology, and Phillip Marcus, at the University of California at Berkeley, are attempting detailed modeling of the Red Spot. They assume that the spot is a large shallow eddy that has developed an enclosed cloud system and is trapped within the zonal wind flow. In an effort to characterize the deep winds, they are studying how such an eddy would interact with its surroundings. The competitive efforts of these investigators have shown that even though their current models are far more simple than they would wish, a detailed model of the problem demands a large computing capacity. Even faster computers will be needed to compute the more realistic models as they are developed. In addition, their work reveals the need to obtain more detailed maps of the atmospheric flow around long-lived eddies and better measurements of the wind variations in some undisturbed region, located at the same latitude, far from the storm system in question. Can new observations be obtained that will yield this information?

Future Atmospheric Research

Although observers become excited by the onset of a new disturbance or formation of a new cloud system, the types of events that are observed on Jupiter are not new. Similar phenomena recur at specific latitudes on time scales of years. Thus researchers want not only high-resolution multicolor images of long-lived features but also a self-consistent set of global observations over as long a period as possible. Such information will be far more valuable if they can obtain simultaneous ultraviolet and infrared spectra of specific cloud features. It would allow investigators to determine which molecules are carried up from lower levels, how much energy emerges from below the clouds, and how the upper atmosphere responds.

The Galileo probe, which will enter Jupiter's atmosphere in December 1995, will yield unique information about the conditions beneath the visible clouds. The level of performance that the investigators get from the Galileo orbiter will determine whether the imaging team will get observations with high enough spatial resolution to obtain the desired wind fields.

Interpretation of Hubble Space Telescope data will add to our knowledge concerning longer term variability. The telescope was designed to be maintained in space for fifteen years. Even though it suffers from mismatched mirrors, smearing problems have been solved with the installation of a new camera. In addition, because the blurring was stable with time, it has been possible to remove it from the images with computer processing techniques that utilize Fourier transforms. For objects like Jupiter, where there is modest contrast between adjacent cloud structure, these deconvolution techniques were quite successful. The resulting data have consistent spatial resolution from image to image that is higher than that obtained with ground-based imaging, thus the time life has already begun and will continue with even higher quality images with the new camera (see plates 4*a* and *b*).

Although competition is great for observing time with the Hubble Space Telescope, this telescope offers my colleagues and me a chance to acquire both imaging data and ultraviolet spectra for more than a decade. Plans for docking with shuttles to allow maintenance and installation of infrared instrumentation in 1997 promise that this much-maligned telescope will provide investigators with a set of data that is complementary to that of the Voyagers, *Galileo,* and the flyby of the Cassini Mission (expected to arrive at Saturn in the first decade of the twenty-first century). These data will be combined with the historical records and observations with new ground-based techniques, especially in the infrared, in an ongoing effort to understand the energy balance and variability within Jupiter's atmosphere. When scientists consider both modeling and observational opportunities, they can expect a considerable increase in the understanding of jovian science within the next decade. In addition to these expected advances, there is still a mother lode of information from the Voyager Mission to tap. The comparative information concerning the zonal circulation of Jupiter, Saturn, Uranus, and Neptune should provide strong constraints for modeling efforts. The wide range of conditions that exist within these four low-density planets will provide additional checks on any general model that is developed to interpret the internal heat source, the magnetic field, and organized flow of the deep interior of these planets.

Cameras that image in visible light as well as infrared instruments will be utilized to develop ideas concerning the circulation of Jupiter's atmosphere. Access to high-speed computers with aggressive approaches for displaying the time-dependent information generated by more complex numerical models will aid in the quest to understand the atmospheres of Jupiter and the other outer planets. Thus we leave the planet with optimism and move on to look at its moons, rings, and magnetosphere.

Plate 1a. A Typical View of Jupiter and Ganymede. Obtained on Jan. 24, 1979, by Voyager 1, this image shows the east-west banding and muted color variation of the belts and zones.

Plate 1b. False-Color View of the Red Spot and White Oval. The gross color enhancement was created to accentuate differences in the reflectivity of the clouds. This plate illustrates the way that color is frequently selected to accentuate structure rather than actual color.

Plate 2a. The Equatorial and Mid-Latitude Region. A brown barge is located above a plume, which is at about 7° north latitude. The southern portion of the Equatorial Zone is more orange than the north. The white turbulent clouds in the lower right lie to the west of the Red Spot.

Plate 2b. The Red Spot and a White Oval. Both spots rotate in a counterclockwise direction about their centers while the white chaotic features rotate in a clockwise sense.

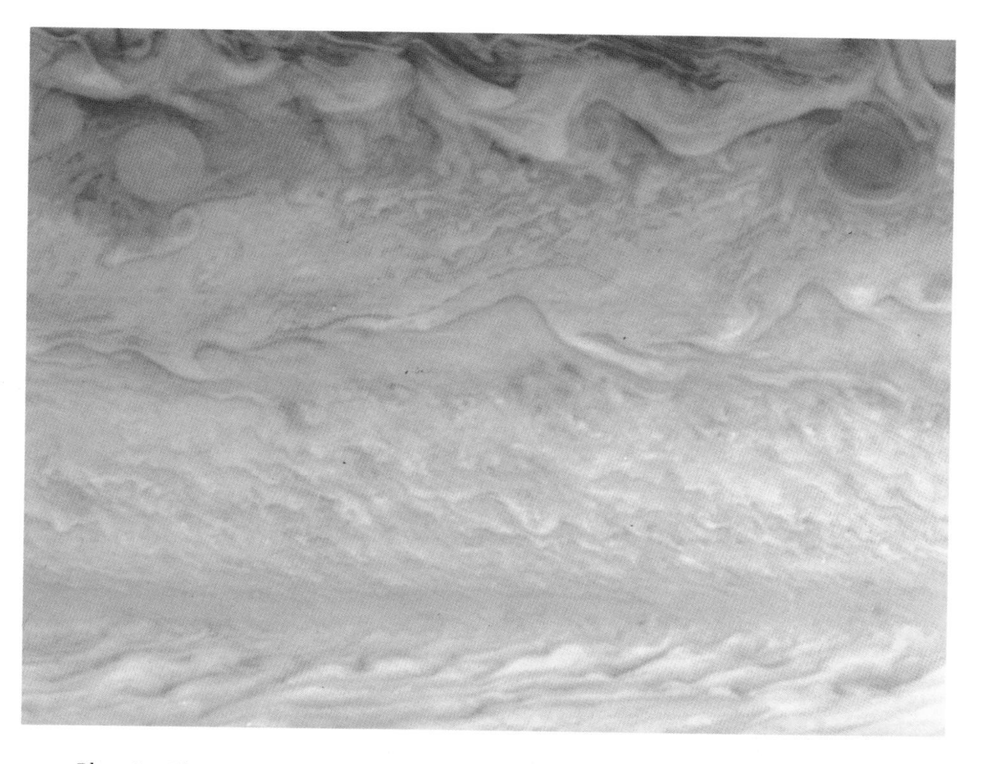

Plate 3a. The Region between 20° and 35° North Latitude. The diagonal pink stripe defines the strong wind jet at 20° north latitude. To the north, a westward flow generates large wavelike structures and brown eddies roll eastward at 32° north latitude.

Plate 3b. Io Silhouetted against the Clouds. To the north of Io, the small orange chevrons mark a strong eastward jet at 7° south latitude. The strongest west wind on the planet generates the large eddies just below Io.

Plate 4a. Hubble Space Telescope Image. In this three-color composite, the blue, green, and red images were preprocessed. Compare this image, which was deconvolved to correct for the telescope's mirror problems, with the Voyager image in plate 4b. (Space Telescope Science Institute, NASA)

Plate 4b. Voyager Image. In this three-color composite, the blue, green, and red images were preprocessed. Compare this image, which was geometrically corrected, with the Hubble Space Telescope image in plate 4a.

Plate 4c. Relative Sizes of the Galilean Satellites.

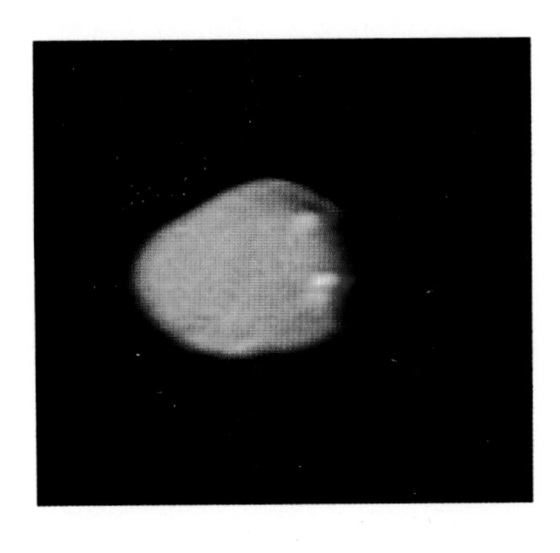

Plate 4d. Amalthea. No dimension on this irregularly shaped satellite spans more than 300 km (200 mi).

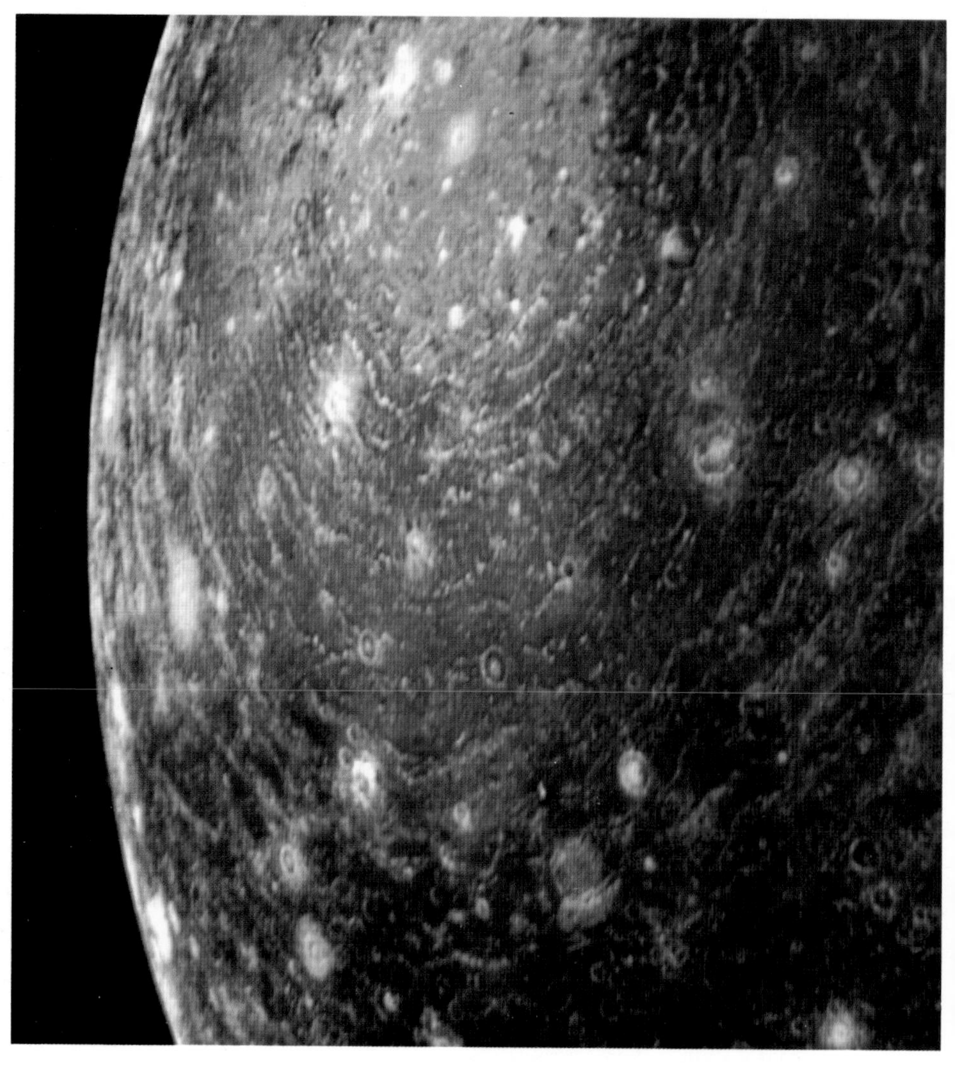

Plate 5a. The Bull's-Eye on Callisto. The concentric structure formed when the crust relaxed inward after the impact of an asteroid or meteor. Note old and recent craters.

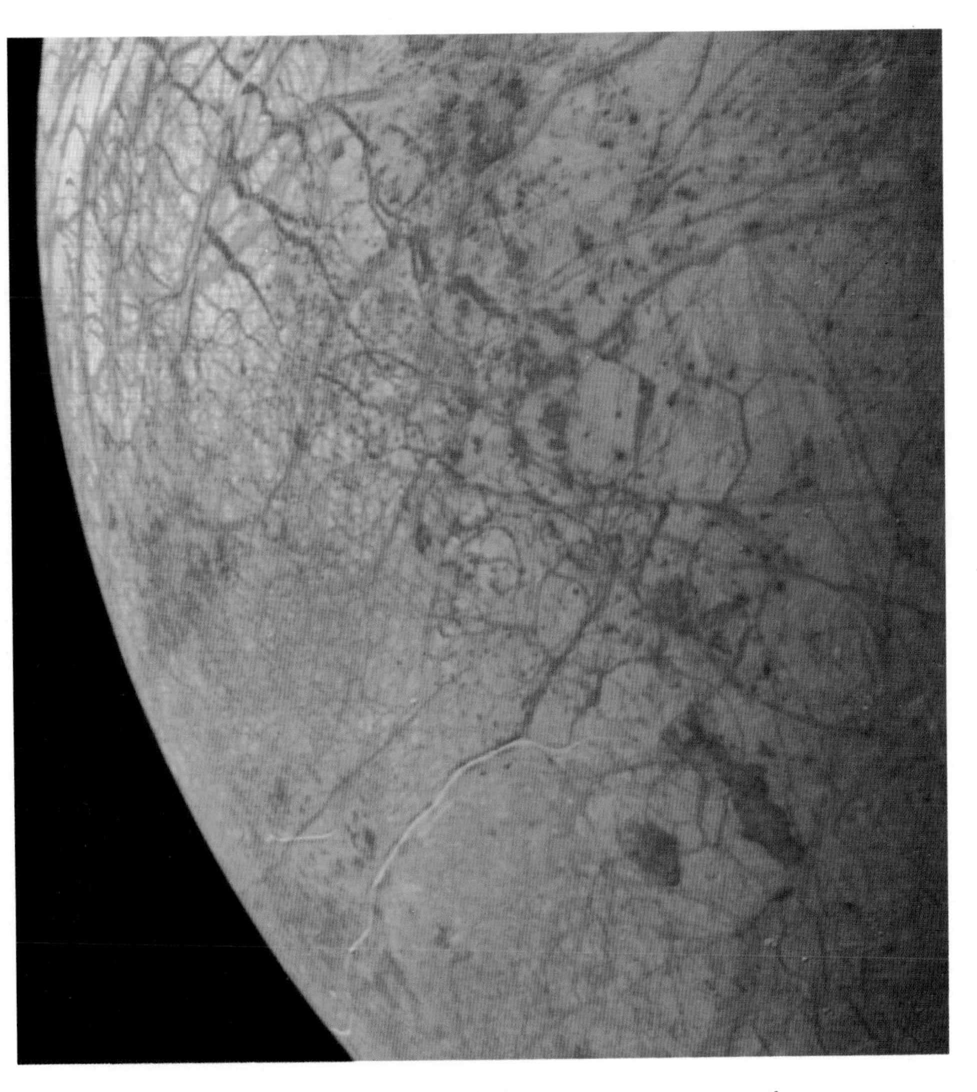

Plate 5b. Europa's Fractured Icy Crust. No impact craters are seen on the surface. This suggests repaving from below at a rate sufficient to eliminate all traces.

Plate 6a. The Galileo Regio on Ganymede. Palimsests, or ghost craters, are visible on the dark, densely cratered surface.

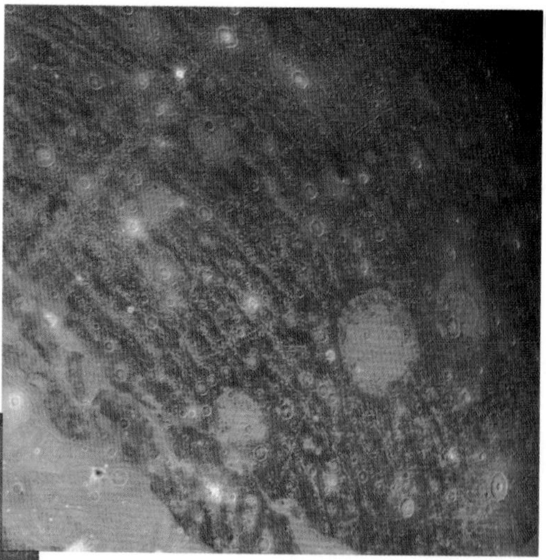

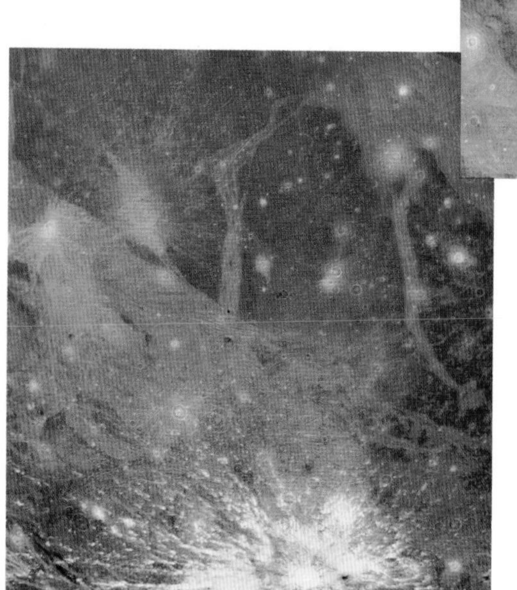

Plate 6b. Fresh Impact Crater on Ganymede. Lighter colored grooved regions are present, overlaid by ices that were ejected from more recent impact sites.

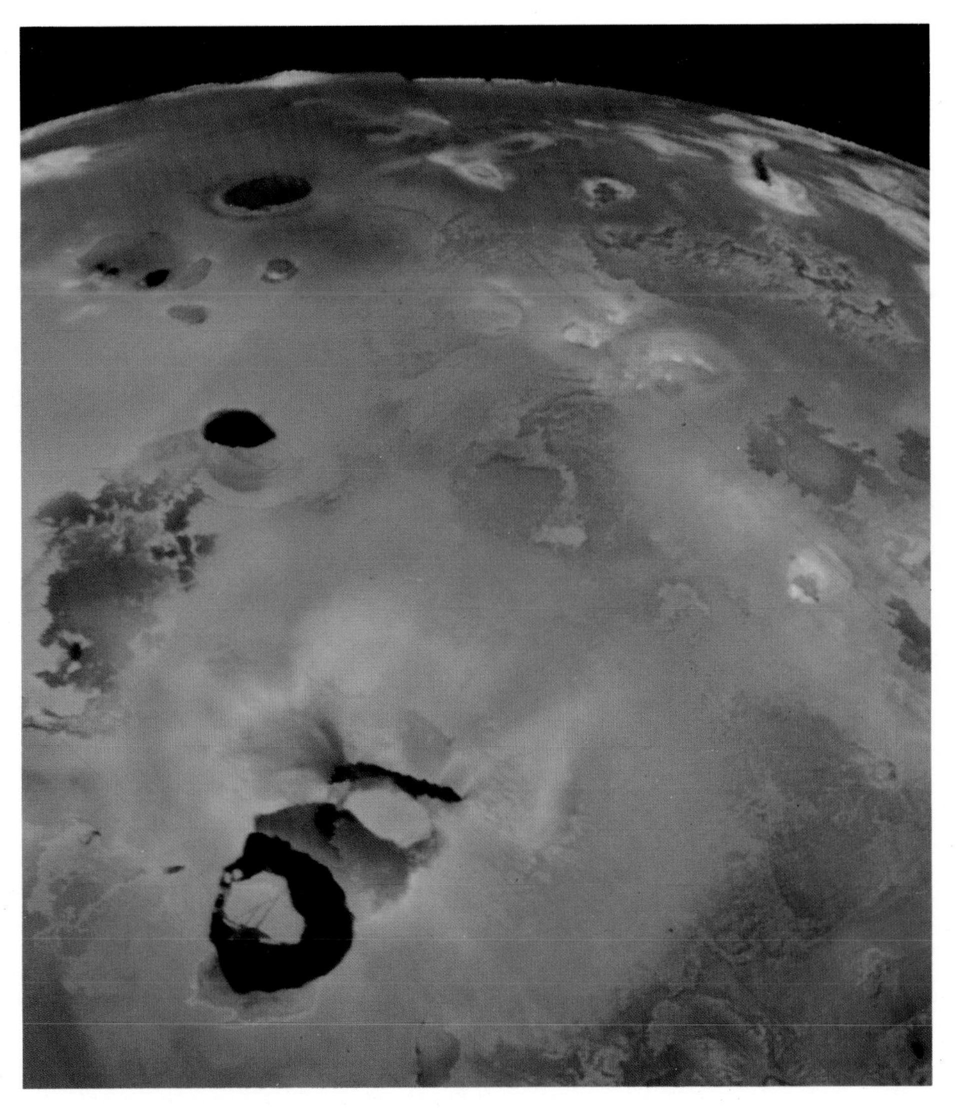

Plate 7a. The Loki Caldera on Io. An effort has been made to reproduce the correct colors, however, the contrast has been enhanced. (Reprocessed by U.S. Geological Survey, Flagstaff, Ariz.; NASA)

Plate 7b. An Eruption of Pele. This image of Io has been processed to enhance the eruption. The mountain catching the sunlight at the bottom of the frame is at 54° south latitude and is 10 km high. (Reprocessed by U.S. Geological Survey, Flagstaff, Ariz.; NASA)

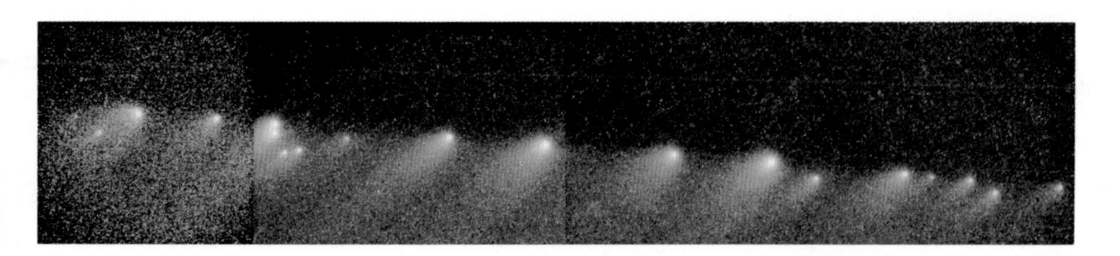

Plate 8a. False Color Hubble Space Telescope Photo of the Individual Shoemaker-Levy 9 Nuclei. Here the color has been adjusted to represent surface brightness grading from dark red to yellow as the brightness increases. (Courtesy of H. Weaver, Space Telescope Science Institute, NASA)

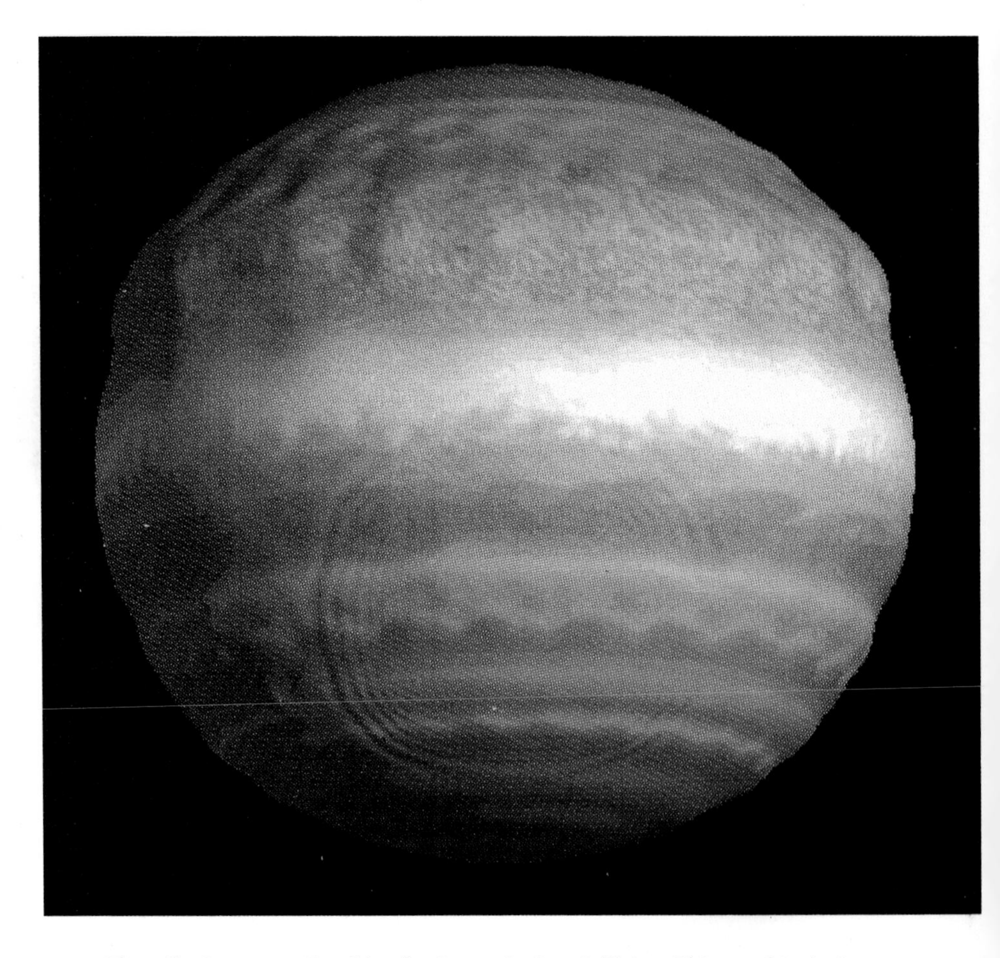

Plate 8b. Computer Graphic of a Comet-Jupiter Collision. This graphic depicts perturbation on pressure 15 hours after a comet (radius = 1 km; density = 1 gm/cc) strikes Jupiter at 37° south latitude, inserting its energy at a depth where the atmospheric pressure = 5 bars. The color and height variations represent pressure variations at this level. How this would modify the overlying clouds is not known. (Courtesy of T. Dowling, Massachusetts Institute of Technology)

Except where noted, these images were provided by JPL, NASA.

♃ PART III

SATELLITES AND RINGS

♃ Chapter 8

Discovery of Jupiter's Satellites

Investigation of the jovian satellites has spanned the entire era of telescopic observations. Although the four large Galilean satellites—Io, Europa, Ganymede, and Callisto—were discovered in 1610, Amalthea, a fifth small satellite orbiting inside Io's orbit was not detected until 1892. The subsequent realization that Jupiter could have many moons stimulated a search for others. Eight more small, distant satellites were detected in the following eight decades, but because of scattered light from Jupiter, discovery of three small satellites near the planet and the thin equatorial ring was delayed until the arrival of the Voyager spacecraft in 1979.

Jupiter's unique system of satellites promises to reveal information about a complex and dynamic history where the moons have been shaped by ongoing physical processes. The Galilean satellites have not only been modified by impacting meteorites but have also been grossly altered by tidal forces generated by mutual interaction as the moons have revolved in the large gravitation field of the massive planet. These satellites display diverse structures, including sulfur volcanoes, glacier fields, grooved terrains, and giant multiringed features (see plate 4).

Although the two Voyager spacecraft obtained considerable data concerning surface features on the Galilean satellites, little is known of the nature of the four small inner satellites or their eight distant siblings. While the small dark-red inner satellites appear to have been highly modified by their environment, the outer satellites seem to be small asteroidal fragments. With this brief summary in mind, let us turn to the history of the discovery of this miniature solar system.

The Galilean Satellites

Galileo Galilei, an Italian scholar, is traditionally credited for the earliest telescopic view of the large satellites. It seems evident that because Jupiter is one of the brightest objects in the sky, anyone fortunate enough to acquire one of the earliest telescopes would have looked at it. But Galileo appears to be the first who realized that the small nearby starlike objects were associated with the planet. He recorded the arrangement of three of the satellites in a published sketch dated January 7, 1610.

At the same time a German scientist, Simon Marius (also Mayr and Mayer), obtained a telescope. In 1611 he announced that he had seen the satellites of Jupiter in December 1609, and in 1614 he published a book to further his claim. In it he referred to observations that he had made as early as November 1609, but his first description of positions was dated December 29, 1609. Galileo was quick to point out that Marius was still using the Julian calendar and that when the discrepancy between the Julian and the Gregorian calendars was applied to Marius' discovery date, it corresponded to January 8, 1610, a day after Galileo's own first observation. In fact, Galileo asserted that Marius' description was taken from his own published accounts of the appearance of Jupiter and the satellites. He accused Marius of plagiarism and a bitter quarrel ensued.

Whether Galileo's accusations were well founded is a topic that science historians still pursue. However, it was Marius who suggested the current names for the four satellites—Io, Europa, Ganymede, and Callisto. Galileo's suggestion that they be given Medicean names in honor of his patron was not widely accepted. Because of the Galileo-Marius controversy, however, for many years the internationally acceptable names for the satellites were Jupiter I, II, III, and IV to avoid crediting a possible plagiarist with the discovery. Just when the current names became fully accepted is not easily ascertained. The records of the International Astronomical Union nomenclature working-group, a committee responsible for assigning names to features on the larger satellites and naming new small outer satellites, provides some insight into the history. Peter Millman (1906–1991), a Canadian planetologist who served as the first chairman of the group, pointed out that there has never been a rival set of names proposed for these satellites. Millman also noted that even though the scientific community had not officially accepted these names until 1975, in a footnote in the 1852 edition of *Outlines of Astronomy,* John Herschel suggested that Galileo be acknowledged as discoverer of the four largest jovian satellites and that the names recommended by Marius be accepted. Therefore, although the four satellites are collectively called the Galilean satellites, their official common names—Io, Europa, Ganymede, and Callisto—have been in general use for some time.

The Other Jovian Satellites

In the waning years of the eighteenth century E. E. Barnard (1857–1928) undertook the task of searching for other satellites of Jupiter. He used the 36-inch Lick telescope, a large refractor located on Mount Hamilton in California, and designed a way to cover the bright disk of the planet as he worked so that he could detect faint nearby objects. On September 9, 1892, Barnard found a small satellite orbiting the planet inside the orbit of Io. He soon ascertained that the period of revolution of the new satellite was slightly less than 12 hours. This satellite was named Amalthea in honor of the goat whose milk was fed to the baby Jupiter. Amalthea was the last of the jovian satellites discovered by direct visual searches, and was designated J V, the fifth known satellite of Jupiter. Astronomers agreed that when others were found they would be sequentially numbered according to the date of their discovery.

The search for fainter satellites required that observers use long exposures on photographic plates. The light from a object too faint to be visible with the naked eye could be registered as a starlike image on a carefully exposed plate. A comparison of three or more of these plates would reveal the systematic motion of a satellite relative to the background stars. In 1904 and 1905 C. D. Perrine (1867–1951) announced that he had found two faint satellites on photographs that had been obtained with the Lick telescope. These faint satellites revolved about the planet beyond the orbit of Callisto, the most remote Galilean satellite. The new satellites were designated J VI and J VII.

Using a 31-inch telescope that had been completed in 1897, astronomers at Greenwich, England, maintained a regular photographic patrol of the sixth and seventh satellites. In 1908 P. Melotte (1880–1961) discovered an eighth satellite while he was studying this series of photographic plates. P. H. Cowell (1870–1949) used observations obtained in 1908 with the Greenwich telescope to compute the orbit of Melotte's satellite. He discovered that it orbited the planet at a more remote distance than any other known satellite and revolved in the opposite direction. Melotte used Cowell's calculations and relocated the satellite the following year, confirming its peculiar retrograde motion.

Work at the Lick Observatory continued, and S. B. Nicholson (1891–1963) announced the discovery of a ninth satellite in 1914. Like J VIII, its orbit lay beyond that of Callisto, and it too revolved about the planet in the opposite direction from the inner satellites. Nicholson continued his observations, and in 1938 he announced the discovery of J X and J XI. However, it was not until 1951 that his efforts were rewarded with the discovery of a fourth outer satellite. With the aid of the 100-inch Mount Wilson telescope, Nicholson became the second person to receive credit for the discovery of four jovian satellites.

Although it had taken Nicholson thirty-seven years to locate four satellites, his work made it apparent that even more satellites might be found if more sensitive observing techniques could be used. But even though observing techniques improved, the thirteenth satellite of Jupiter was not discovered until 1975. Charles Kowal was using a 48-inch Palomar Schmidt telescopic camera, an instrument well suited to carrying out detailed photographic surveys, when he located the faint moon. Although he carried out an extensive search, his efforts revealed no other satellites. The thirteenth was the last of the satellites to be discovered by earth-based observers.

Cameras on board a Voyager spacecraft approaching Jupiter could be pointed so that overlapping images could be used to map regions near the planet. Due to the reduced scattered light in the airless environment, observations could be made closer to the bright disk of the planet. In this way the region around the planet could be systematically searched for unknown satellites. Three other small inner satellites were found. Depending on the length of the exposure, the motion of the satellites about the planet caused them to appear as either small bright dots or as short bright trails on the Voyager images. Proof that these features were satellites was obtained by selecting the suspected feature in an image and determining the distance to the center of Jupiter. Then, assuming the satellite was in a circular orbit, the rate with which the satellite would revolve around the planet was computed and the location of the satellite as a function of time was calculated. These predictions were used to search the other images obtained with the Voyager cameras. Additional sightings of the same object near the predicted locations would confirm that it was a satellite. J XIV was located by D. C. Jewitt and G. E. Danielson, from the California Institute of Technology, and J XV and J XVI were identified by Steven Synnott, from the Jet Propulsion Laboratory.

J XIV and J XVI have diameters of 20 to 40 km and orbit the planet in a little more than 7 hours. J XV is also small, but this satellite revolves about Jupiter in about 16 hours in a slightly larger orbit that lies between Io and Amalthea.

The inclusion of these three new satellites results in a current membership of the family of jovian satellites that consists of the four large Galilean satellites and at least twelve small satellites. The Galilean satellites orbit the planet in nearly circular orbits in the equatorial plane of the planet. Their orbits are located within 6 to 26 planetary radii of the center of the planet and they revolve about it in 1.8 to 16.7 days. The twelve smaller satellites can be considered as members of three different groups. The innermost group of moonlets, consisting of Amalthea and the three new satellites observed by the Voyager cameras, orbit the planet within the orbit of Io at distances smaller than 3.5 planetary radii. Possibly they are fragments from a larger body that was disrupted in the past. The International Astronomical Union has accepted the mythological names Andrastea, Thebe, and Metis for J XIV, J XV, and J XVI, respectively.

The other eight form two sets of outer satellites. One group consists of four small satellites that revolve about the planet in elliptical orbits at average distances of 155.5 to 164.5 planetary

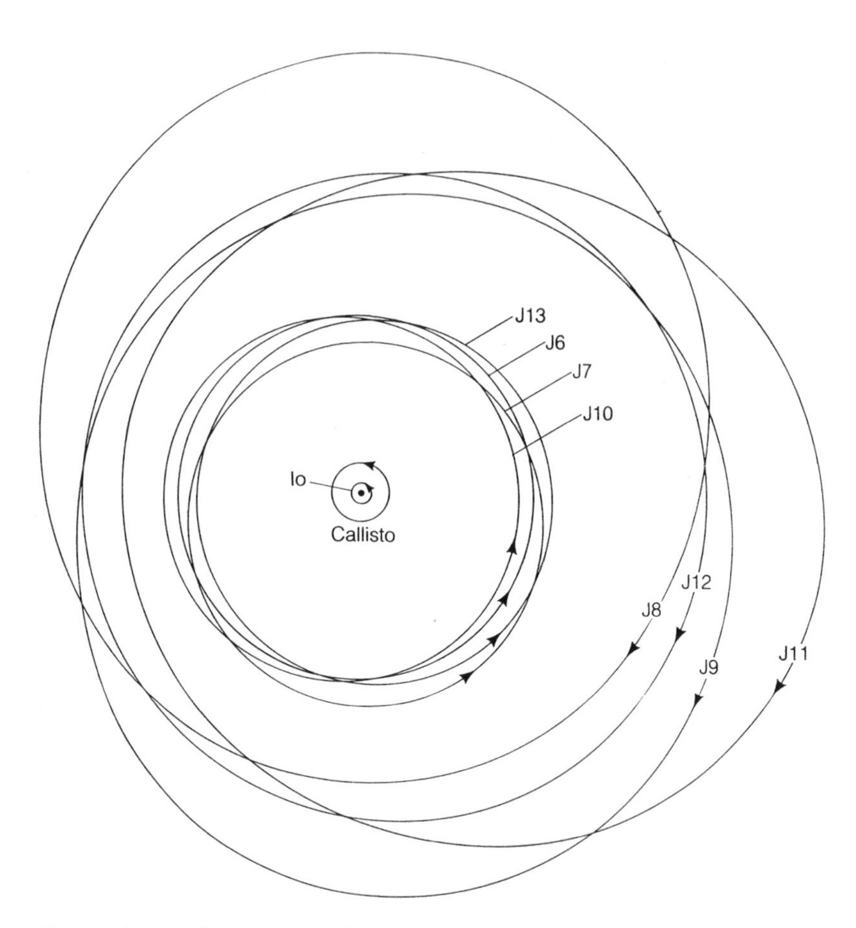

Figure 18. Scale Drawing of the Orbits of the Outer Jovian Satellites. The elongation of the outer orbits is apparent. (Courtesy of W. Hartmann. Reprinted from W. K. Hartmann, *Moons and Planets,* 3d ed. Belmont, Calif.: Wadsworth Publishing Co., 1993, p. 147)

radii and require 239 to 260 days to complete a revolution. Because of their eccentric orbits, the distance of closest approach and the most distant departure of these satellites from the planet varies greatly, ranging from 130 to 190 jovian radii. These satellites of Jupiter were initially assigned only Roman numerals because Nicholson declined to name the four that he had found. Finally, after the problem of the names of the Galilean satellites had been solved, astronomers agreed on names ending

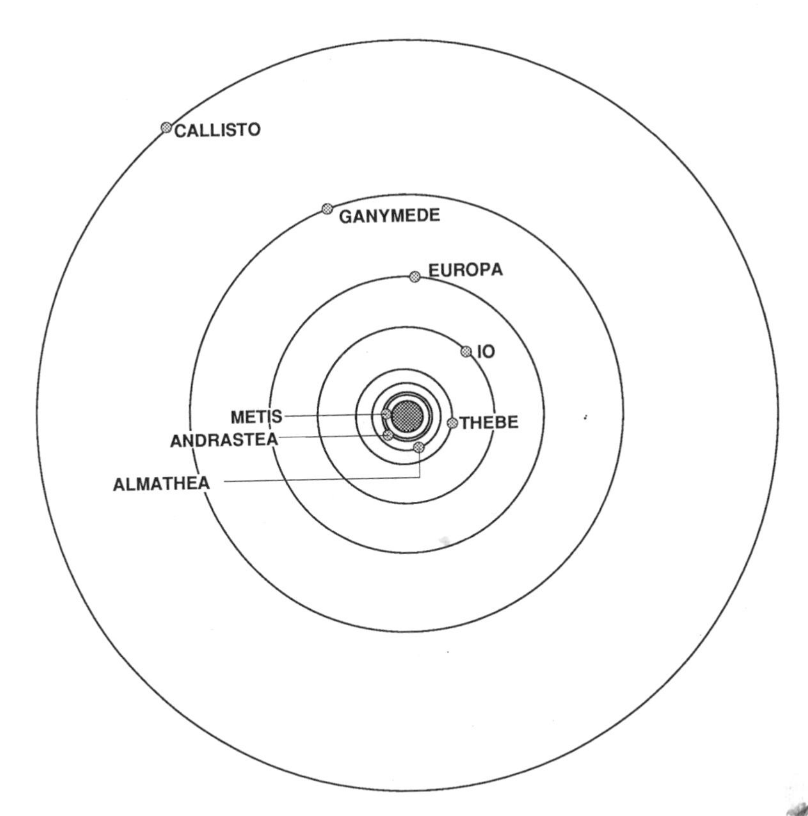

Figure 19. An Expanded View of the Inner Portion of the Jovian System. Together this diagram and figure 18 illustrate the wide range in the size of the satellite orbits.

with the letter *a* for this group. Thus the members of the "a" satellites are Himalia (J VI), Elara (J VII), Lysithea (J X), and Leda (J XIII). The orbits of these satellites are not only elliptical, they are also tipped 25° to 30° relative to the planetary equator. The apparent brightness and the manner with which these satellites reflect ultraviolet, blue, and visual sunlight indicate that they are small asteroid-like bodies with diameters ranging from 10 to 100 km. The similarity of the orbits of these four satellites suggests that they may have had a common origin and could be fragments of a larger body that was destroyed by a collision.

The members of the most remote family of four satellites orbit the planet at distances ranging from 240 to 320 planetary radii. These are the "e" satellites: Pasiphae (J VIII), Sinope (J IX), Carme (J XI), and Ananke (J XII). The individual satellites revolve about the planet in orbits that are inclined by 16° to 30° relative to Jupiter's equator; their periods of revolution are 231 to 758 days. Their orbits are quite elliptical, and over each revolution their distance from the planet varies from 17 to 38 percent of the average distance. The most unusual thing about these 10 to 20 km sized objects, however, is that they revolve about Jupiter in the direction opposite to that of the other satellites and opposite to the direction that the planet rotates on its axis. The fact that these four outer jovian satellites are located in similar orbits and share this peculiar retrograde motion leads to the conjecture that they too are trapped remnants of a larger body that was involved in a destructive collision near Jupiter.

In summary, the jovian ensemble of known satellites consists of four groups, each composed of four satellites: the small inner ones, the Galilean satellites, the distant small group in direct revolution about the planet, and the remote group that are in reverse, or retrograde, orbits. Although thirteen of these satellites were known and positions could be predicted for times when a spacecraft would arrive at Jupiter, Amalthea was the only small satellite that could be effectively observed by the Voyager 1 or 2 spacecraft. These observations were folded in with those of the Galilean satellites and the jovian cloud deck, and will be discussed in chapter 11.

♃ Chapter 9

Pre-Voyager Knowledge of the Galilean Satellites

As the Voyager team attempted to plan the observing sequences for the Jupiter encounters with the two Voyager spacecraft, they became painfully aware of the uniqueness of data sets that they could acquire. They knew that space probes would not visit Jupiter frequently and that the period of time over which they could obtain high-resolution observations was seriously limited. Every effort was made to use pre-Voyager knowledge and ideas of how the solar system formed to optimize the yield from the mission. A major goal of the participating geologists was to generate universally acceptable maps of the surface features of the satellites. A review of the early planning and the extent of the pre-Voyager understanding of the Galilean satellites will help us to appreciate the differences among these moons.

Variations in Surface Brightness

Even though observers could not resolve features on the satellites with earth-based telescopes, they could measure the brightness of the entire unresolved disk through a series of filters that were transparent to selected colors of light. As early as the mid-1920s photometric studies of this type revealed that when the reflectivity of the satellites was compared to the brightness of nearby background stars, their brightness could vary with time by as much as 40 percent. It was soon established that the variation was cyclic and that the period of variability for each satellite was equal to its period of revolution about Jupiter.

These observations were recognized as evidence that these

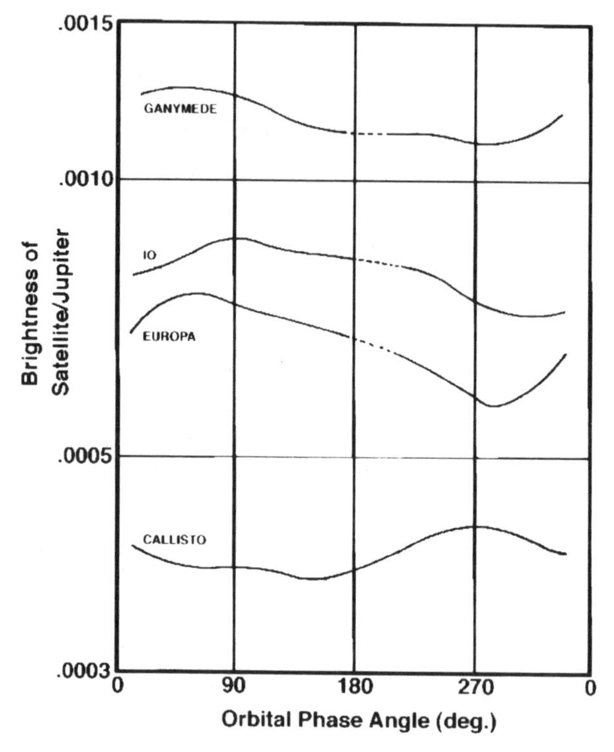

Figure 20. Variation of Surface Reflectivity Relative to Jupiter's Brightness. Zero longitude is at the subplanet point, with the longitude increasing through the leading hemisphere. (Courtesy Dale Cruikshank, Ames Research Center, NASA)

satellites, like our moon, rotate on their axes at the same rate that they revolve around the planet. When this type of rotation occurs, one side of the satellite continuously faces the mother planet and the satellite is in synchronous rotation. An earth-based observer monitoring one of these synchronously rotating satellites would see a configuration such that when the satellite is viewed at a maximum eastward separation, or elongation, from the planet, it would be moving away, and when viewed to the west of the planet, it would be approaching. Therefore the observer would alternately view the trailing and leading hemispheres of the body. When this geometry is considered, it becomes clear that the photometric variations of a given satellite indicate that the reflectivity of the leading and following hemispheres is different. On Ganymede, Io, and Europa the leading side is brighter, but on Callisto the brightest region is on the trailing side.

Astronomers could not deduce the cause of these variations without information about the small-scale surface structure; thus it was apparent that the Voyager team should strive to obtain as much surface coverage as possible with the cameras. Because of the rotational lock of the satellites, for a given satellite it was desirable that one Voyager spacecraft arrive when it was on the near side of Jupiter so that the back side would be illuminated by the sun. The other spacecraft should arrive when the satellite was on the far side, when the planet-facing side would be illuminated. The choice of the right trajectories to accomplish this with only two spacecraft and four satellites, all revolving about the planet at different rates, was certainly not trivial.

The evidence that the satellites of Jupiter were in synchronous rotation had additional ramifications. Investigators knew that the tidal forces that the earth had exerted on the moon had created stresses and slowed the rate of rotation until the moon became tidally locked, with one face always turned toward the earth. Comparison of the surface features on the far side with those on the earth-facing side of the moon indicates that this kind of tidal force played a decisive role in the formation of the large maria, or seas, of dark lava that dominate the earth-facing side. This awareness led investigators to expect that Io and Europa, the inner satellites, had been subjected to more extreme tidal modification than Ganymede and Callisto, the outer satellites. They considered available information that indicated this was the case. Two different known properties, spectral characteristics of the surface and density variation as a function of distance from the planet, substantiated these ideas.

Spectra of the Satellite Surfaces

Long before the Voyager Mission, astronomers had convincing evidence that the surfaces of the four satellites had considerably different reflective properties. When the measured brightness of Io was compared to the size of the observable disk, it was determined that the satellite was nearly as reflective as fresh white snow but was also surprisingly red relative to Jupiter or the other satellites. Europa reflected almost as much sunlight but was not red, while Ganymede and Callisto reflected a smaller percentage of the incident light and showed little color dependence. The reflectivities of the two latter satellites were characteristic of varying mixtures of ice and rock. These data were obtained with filters that transmitted a wide range of colors, and although they indicated that the surfaces of the satellites were quite distinguishable from each other, the data could not be used to specify the composition of the reflecting surfaces.

To determine the chemical composition, detailed information concerning the manner in which the satellites reflect different colors, or wavelengths of light, is needed. This can be achieved either by constructing a series of filters that transmit narrow color ranges or by employing a spectrograph. Depending on how the spectrograph is designed to disperse the incoming light into its constituent colors, it will measure in detail the amount of sunlight reflected throughout a range of colors, or wavelengths. When a spectrum is obtained for a satellite and the reflected light is compared to a standard solar spectrum, the difference between the two would be due to the absorption properties of the surface of the individual satellite.

When ground-based infrared spectra of the whole visible disk of each satellite were compared to laboratory spectra of known substances, it was obvious that the spectra of Europa and Ganymede contained a strong signature of absorption due to water ice. This was weak or missing in the spectra of Io and Callisto. The reflection spectra of Callisto appeared to be related to that of a dark rocky surface, similar to what would be expected if the surface were covered with asteroidal material. The spectrum of Io was different: it was dark in the blue region and highly reflective in the red and infrared regions. Comparisons with laboratory spectra indicated that the satellite could be covered with sulfur or common salts such as sodium chloride (table salt) or sodium sulfate (Epsom salt). This idea had been reinforced when ground-based observers detected emission from a sodium cloud that was related to Io's orbit. Presumably the material in this cloud had escaped from the satellite.

Plans to obtain Voyager data to explore these questions included a combination of observations utilizing ultraviolet and visible filters with the Voyager cameras and using the infrared spectrograph. Observing sequences that would allow mapping of the satellites and surveying Io's orbit in the region that contains the sodium cloud were planned. The fact that the spacecraft would pass close enough to the satellites to obtain spectra as a function of position on the satellite promised that many of these questions could be solved and that the data would reveal information concerning the history of these bodies.

Density of the Satellites

Before the exploration of space astronomers were aware that even though the two outer satellites, Ganymede and Callisto, were more massive, the average densities of Io and Europa were greater. Astronomers had combined a series of observations that had been acquired over a long period of time to obtain reasonable estimates of the diameters and masses of each satellite. Because the distance to the Galilean satellites was well known, an observer could crudely determine their radii if the apparent angular sizes of the satellites could be measured. This was done at times when the smearing due to motions within the earth's atmosphere was small. The masses could be determined by combining many measurements of the location of the satellites relative to Jupiter. Although the planet is so massive that it almost totally dominates the system and the individual perturbations that the satellites exert on each other's orbits are small, the effects are cumulative. Thus over a period of time it is possible to separate the effects of each satellite. Even in the late eighteenth century astronomers were able to deduce the masses of the satellites.

If the satellite is assumed to be spherical and its mass and diameter are known, the volume can be calculated and the average density determined by dividing the mass by the volume. Calculations of this sort revealed that the densities of Io and

Europa were about three and a half times the density of water and similar to the density of our moon. Surprisingly, the densities of the large outer satellites, Ganymede and Callisto, were less than twice the density of water. This low density suggested that these bodies must contain a large component of low-density ices.

At Jupiter's distance from the sun and in a hydrogen-rich environment, most of the nitrogen and carbon would have formed gaseous ammonia and methane, while the oxygen would have combined with the hydrogen, causing water ice to form. At the expected local temperatures, dirty lumps of water ice containing meteoric grains probably accreted onto one another to form these ice-rich satellites. So the question was not why the densities of the outer satellites were low, but why the densities of the inner satellites were high.

Radii of the Satellites

Because of the stringent time constraint imposed by a mission that involves a flyby of the spacecraft, it was apparent from the earliest planning stages of the Voyager Mission that it was desirable to determine the radii of the satellites as accurately as possible. This would allow the Voyager team to use a minimum number of camera frames to map each satellite and improve the planning of the spectroscopic observations. Thus in the early 1970s NASA funded investigators to attack the problem with ground-based telescopes.

Observers knew that better measurements of the diameters of the satellites could be made with modern high-speed photometers if they could observe a satellite when it occulted a star. To obtain this information, an astronomer would measure the total amount of light arriving from the star and the satellite. The goal of the observation is to detect a decrease in brightness and to measure its duration. Because the star is at such a great distance, the duration of the occultation is due to the known rate of motion of the satellite and the unknown diameter. The position of Jupiter against the background stars and the motion of the satellites about the planet can be calculated, and predictions of the time and location on earth where such events can be viewed can be made well in advance.

Although favorable opportunities are not frequent, observations were obtained in the early 1970s when Io and Ganymede occulted bright stars. In 1971 Io occulted a bright star in the constellation of Scorpius, and several groups, headed by G. E. Taylor, participated in attempts to observe the event. They reported a radius of 1820 km with a measurement uncertainty of ±10 km. An occultation of a fainter star by Ganymede in 1972 resulted in R. W. Carlson, along with many of the participants in the Io event, reporting a radius of 2635 ±25 km.

Efforts to obtain occultations involving Europa and Callisto were fruitless; another approach was used to determine the diameter of Europa. Once every six years the line of sight for earth-based observers is perpendicular to Jupiter's rotational axis and an observer sees the Galilean satellites in its equatorial plane. As they revolve around the planet, they occult one another. If the diameter of one satellite is well known, an observer can use Kepler's laws and the occultation data to derive the diameter of the other. This requires careful timing of the duration of the events. K. Aksnes and F. A. Franklin reported a radius for Europa of 1500 ±100 km based on analysis of 1973–1974 events that involved Io and Europa. Therefore, by the time that the Voyager imaging team completed the scheduling of the satellites, only Callisto's radius had not been remeasured by these more accurate methods. The radius of Callisto was expected to be 2500 ±150 km, based on direct measurements of the diameter of its apparent disk.

The improved radii were also used to compute the volume of the satellites, and the results substantiated the fact that the density of the satellites decreases with distance from the planet. Furthermore, when Pioneer 10 and 11 passed through the jovian system in 1973 and 1974, the precision to which the masses of the satellites were known improved. Because the masses of the Pioneer spacecraft were known, it was possible to refine the masses of some of the satellites based on the forces that the two spacecraft experienced as they flew by. Therefore, by the time of arrival of the Voyager spacecraft in 1979, the densities of the satellites were known to decrease outward from the planet, with the mean densities ranging from 3.5 gm/cc to less than 2 gm/cc from Io to Callisto.

These results indicated that the inner satellites either had formed from more dense material or that the tidal forces had created a situation such that the lighter elements had escaped. The latter was easier to explain. The most generally accepted model for the system predicts that collisions between particles that are orbiting the planet in inclined orbits and the effect of the oblate planet itself will cause the satellites to settle into the equatorial plane. This model also requires that the orbits of widely separated satellites become more circular with time. However, the three inner satellites are located at distances such that Europa's period of revolution about the planet is twice that of Io and Ganymede's period is four times that of Io. The recurring alignments of the satellites prevent their orbits from becoming perfectly circular. Although the current configuration may have evolved from one that was considerably different, at the present time each satellite imposes a systematic variation on the others, maintaining slight deviations from circular orbits. Because of this, as each revolves around Jupiter, its distance from the planet varies slightly; as this occurs, the size of the force that creates a tidal bulge increases or decreases and the satellite is forced to adjust. The resulting deformation will generate stresses within the bodies, resulting in frictional heating that will depend on the tidal forces and the internal structure of the satellite. These forces are due to the gravitational field generated by Jupiter's large mass, and they will increase rapidly as the distance to the center of the planet decreases. As a result of this ongoing action, the innermost satellite, Io, should suffer the most alteration, while Callisto, the outermost satellite, should be the least altered and have the most primitive surface.

The manner in which the three inner Galilean satellites revolve around Jupiter not only creates the driving forces that modify their structure and composition, it also ultimately determined the image quality and total area of the satellites that Voyager could map. The periods of revolution of Europa and Ganymede, two and four times that of Io, respectively, and the placement of the satellites around the equatorial plane are such that these three inner satellites will not group together on one side of the planet. Therefore it was not possible to select a point in time where a single spacecraft could have a close encounter with all four satellites. Instead, the near approaches to the satellites had to be judiciously distributed between the two encounters to optimize the total area and spatial resolution that could be observed.

Guidelines for Naming Surface Features

As the team planned for the construction of detailed maps, it became apparent that standard nomenclature would be needed. Team members who were also members of the Planetary Commission of the International Astronomical Union (IAU) took the lead in establishing a system that would provide an orderly scheme for assigning names to features. At the time the IAU Working Group for Planetary Nomenclature had already had a taste of the difficulties associated with generating universal maps of the moon and inner planets. It met in Moscow in 1975 and agreed to use names of characters from world-wide mythologies. The universality of these tales would eliminate the implication that any country was establishing a territorial claim. The group also agreed to continue the use of Latin terms to designate geological features. The Latin term would describe the appearance of the feature on the surface, not indicate how it had formed. Features on Callisto would be named after characters from the sagas of the Norsemen; on Ganymede, they would be of Middle Eastern derivation, based on Sumerian, Babylonian, Assyrian, and Egyptian myths. Characters from the original myths of Callisto and Ganymede would also be acceptable. Structures on Europa and Io would be named for places and characters derived from the legends associated with the maidens for whom the satellites were named. Lists of names were prepared in advance and circulated within the IAU committee, and the authenticity of the mythological sources was documented. This groundwork provided a firm basis for generating maps of the surfaces of the satellites, and because of the efforts of dedicated individuals such as the late Harold Masursky (1923–1990) from

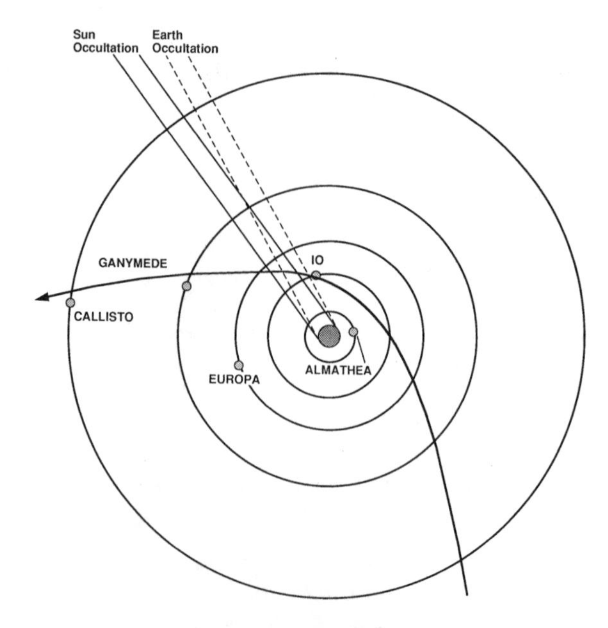

Figure 21. Path of Voyager 1 among the Jovian Satellites. This encounter yielded only a distant approach to Europa. (Adapted from NASA SP-439)

the U.S. Geological Survey, universally accepted maps have been constructed.

The IAU proposals were followed for Callisto, Ganymede, and Europa. However, after arriving at Io and discovering the volcanic surface, the investigators, with the blessing of the IAU, chose to name surface features for gods of fire, the sun, thunder, and lightning. For example, two of the large volcanoes were named for Pele, the Hawaiian fire god, and for Prometheus, the Greek Titan who stole fire from the gods.

With agreements on nomenclature reached, Voyager team members turned their efforts toward planning every detailed step of the observations for the two encounters.

Arrival of the Spacecraft

Voyager 1 arrived at Jupiter in early March 1979, when the four Galilean satellites were positioned such that only a distant pass of

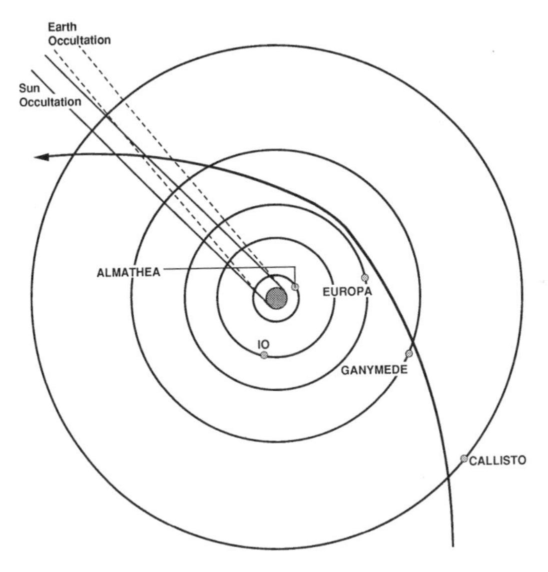

Figure 22. Path of Voyager 2 among the Jovian Satellites. This time Io is observed from a distance. (Adapted from NASA SP-439)

Europa was possible and the encounters with Io, Ganymede, and Callisto occurred after closest approach to the planet. The situation was reversed with Voyager 2, in July 1979. This time the spacecraft passed by Callisto, Ganymede, and Europa before encountering the planet and failed to achieve a close pass by Io. This schedule allowed Voyager 1 to view the hemispheres that constantly face the planet and Voyager 2 to see the backsides, or hemispheres that were facing away from the planet, while they were illuminated. It was hoped that the resulting high-resolution data would answer many of the questions that had been raised by the low-resolution earth-based observations.

The planned Voyager sequences were successfully executed and systematic mapping and interpretation of the data is still under way. The results that were obtained for Callisto and Ganymede are presented in chapter 10. Information concerning Europa and Io and a review of the limited observations of Amalthea are included in chapter 11.

♃ Chapter 10

Callisto and Ganymede

During the Voyager encounters in 1979 some of the tantalizing questions raised by earlier planetary studies were answered while others remained to challenge investigators. The unprecedented spatial resolution of both imaging and infrared data led to many discoveries and, not surprisingly, many new questions. Inspection of Galilean satellite images immediately shows that Callisto and Ganymede have dark cratered surfaces. Respectively about 4.5 and 2.5 times farther from Jupiter than Io, the innermost of Jupiter's four large satellites, the surface modification on these two has occurred more slowly and they have suffered less alteration than either of the others. Only 1.6 times more distant from Jupiter than Io, Europa is covered with an icy crust that is mobile enough to destroy evidence of meteoric infall onto the surface. Io's surface is even more plastic. Thus a logical way to look at these evolved bodies is to describe the outer two and then proceed inward toward Jupiter.

During this discussion we will use a standard designation of latitude and longitude on these tidally locked satellites. If you place the curled fingers of your right hand in the direction that the satellite rotates on its axis and revolves about the planet, your thumb will point north. From this you can establish a mental picture of the northern and southern hemispheres or latitudes relative to a satellite's motion. The point corresponding to 0° latitude and 0° longitude is assigned to the point that is closest to the planet and longitude is defined in a manner such that the 90° meridian crosses the equator at the point that leads the satellite in its motion around Jupiter. Thus 180° is opposite the subplanet point and 270° is on the trailing face.

The Bulk Characteristics of Callisto and Ganymede

Callisto, the outermost of the four largest jovian satellites, revolves about Jupiter in 16.67 days at a distance of about 15 planetary radii. The satellite has a diameter of 4800 km, which is just slightly smaller than that of the planet Mercury. The average density of this satellite is 1.8 gm/cc, compared to a density of 1 gm/cc for water and 3 to 5 gm/cc for ordinary rock. Such low density excludes the possibility that the bulk of this satellite can be rock. Because it formed in a deep-freeze environment at a distance five times farther from the sun than the earth, it is not surprising that water ice, composed of the commonly occurring deep-space elements hydrogen and oxygen, may have been a major constituent of the material that condensed to form the body. The low reflectivity and infrared spectra derived from ground-based observations are consistent with a relatively inactive body in an atmosphere-free environment. The old surface would be covered with meteoric dust and cratered by larger meteors.

In comparison, Ganymede is the largest satellite in the solar system. With a diameter of 5260 km, it is considerably larger than the planet Mercury. With a period of 7.16 days, it revolves about Jupiter in a nearly circular equatorial orbit at a distance of 9.5 planetary radii from the center of the planet. Infrared spectra show that, unlike Callisto, a large amount of water ice is present on Ganymede's surface. Nevertheless, the fact that Ganymede reflects only 44 percent of the light that is incident on its surface indicates that some portion of the surface is covered with dark material. Based on its reflective properties, it was generally assumed prior to the Voyager flyby that Ganymede was covered with dark meteoritic material, with fresh white ice exposed in areas where larger meteors had impacted.

The average density of Ganymede is 1.9 gm/cc, which indicates that, like Callisto, a major constituent of the material from which the satellite formed must have been ice. Pre-Voyager uncertainties concerning Ganymede were related to the question of whether the icy crust of such a large body could support craters for an extended period of time. The cratering impact rate as a function of time was unknown, and the extent of tidal flexing of the satellite was not defined. Because of these unknowns, predictions ranged from a crust that was saturated with ancient impact craters to a surface that was so fluid that it was not capable of preserving a record of even relatively recent impacts.

With this information in hand, the original imaging science team, which consisted of Bradford Smith, Geoffrey Briggs, Allen F. Cook, G. E. Danielson, Jr., Merton Davies, Garry E. Hunt, Harold Masursky, Tobias Owen, Carl Sagan, Lawrence Soderblom, and V. E. Suomi, assisted by additional invited investigators and support staff, selected the observations that would best reveal the surface structure of the satellites.

The Surface of Callisto

After transmitting spectacular images of the planet and executing an intense observing schedule during closest approach to Jupiter on March 5, 1979, the Voyager 1 spacecraft passed near Callisto as it moved out of the jovian system to continue on its way to encounter Saturn in November of 1980. The spacecraft passed within 126,000 km of Callisto, a distance that limited the best resolution of surface features with the narrow-angle camera to 2.3 km. The Voyager 2 spacecraft arrived at Jupiter when Callisto was on the sunward side of the planet and provided a view of the side of the satellite not seen by Voyager 1. This allowed mapping of most of its surface. However, the closest approach to Callisto of Voyager 2 was 220,000 km, yielding images of the satellite's surface on the side facing away from the planet that were less sharp than those on the facing side. At the distance of closest approach features larger than 8 km in diameter would span more than one pixel of the narrow-angle camera.

The Voyager cameras revealed a surface that was dark, but violet, blue, green, and orange filtered images revealed that the darker regions were more red than the lighter areas. The most striking feature on the surface was a large "bull's-eye" structure (see plate 5*a*). This feature showed a brighter central region 600 km in diameter and centered at 10° north latitude and 55° longitude. The central region was surrounded by concentric rings separated by 50 to 200 km and extending outward 2000 km, or five-sixth of Callisto's radius. In keeping with pre-encounter agreements, this feature has been named Valhalla, for the Norse warriors' heaven. A similar smaller feature, located at 30° north latitude and 140° longitude has been named Asgard, after the heavenly residence of the Norse gods. At least six other smaller ringed structures have been identified in the Voyager data.

Although Mare Orientale on the Moon and Caloris Basin on Mercury appear to be large multiple-ringed impact craters, the concentric ridges surrounding them are few in number and vertically mountainous. On Callisto the reduced vertical structure and multiplicity of the rings may be due to the more fluid properties of the crust. The density of craters in the bright central area of Valhalla is about one-third as great as the average density on the surface of the satellite, indicating this feature is younger than the heavily cratered regions. In the outer ringed area surrounding this feature many of the ridges cut through craters and the crater density appears normal, indicating that the ridges formed after the craters. Other craters overlay the ridges, providing evidence that the bombardment continued after the formation of Valhalla.

It appears that Callisto's crust reacts differently to a large impact than that of our moon or Mercury. In all cases the kinetic energy, or energy of motion, must be deposited instantaneously at the impact site, and although results from laboratory experiments indicate that thick layers of ice react to impacts in a manner similar to rock, the post-impact recoveries would differ. William McKinnon, from Washington University in St. Louis, proposed that the concentric rings formed during the recovery phase. A large impact such as this one would insert energy well below the surface. If the outer part of the satellite consists of an icy mantle that is overlaid by a lithosphere, or surface consisting of a layer that is mostly rock, both layers would flow radially outward from the point of impact. During the recovery phase the lower, more plastic mantle would flow back into the site of the impact more readily. The weight of the lithosphere, however, would prevent easy slippage along the interface, and traction forces would develop that were directed radially inward. As the mantle flowed in to fill up the impact basin, the upper crust would crack and units would be dragged along by the moving mantle, forming concentric expansional faults, or grabens.

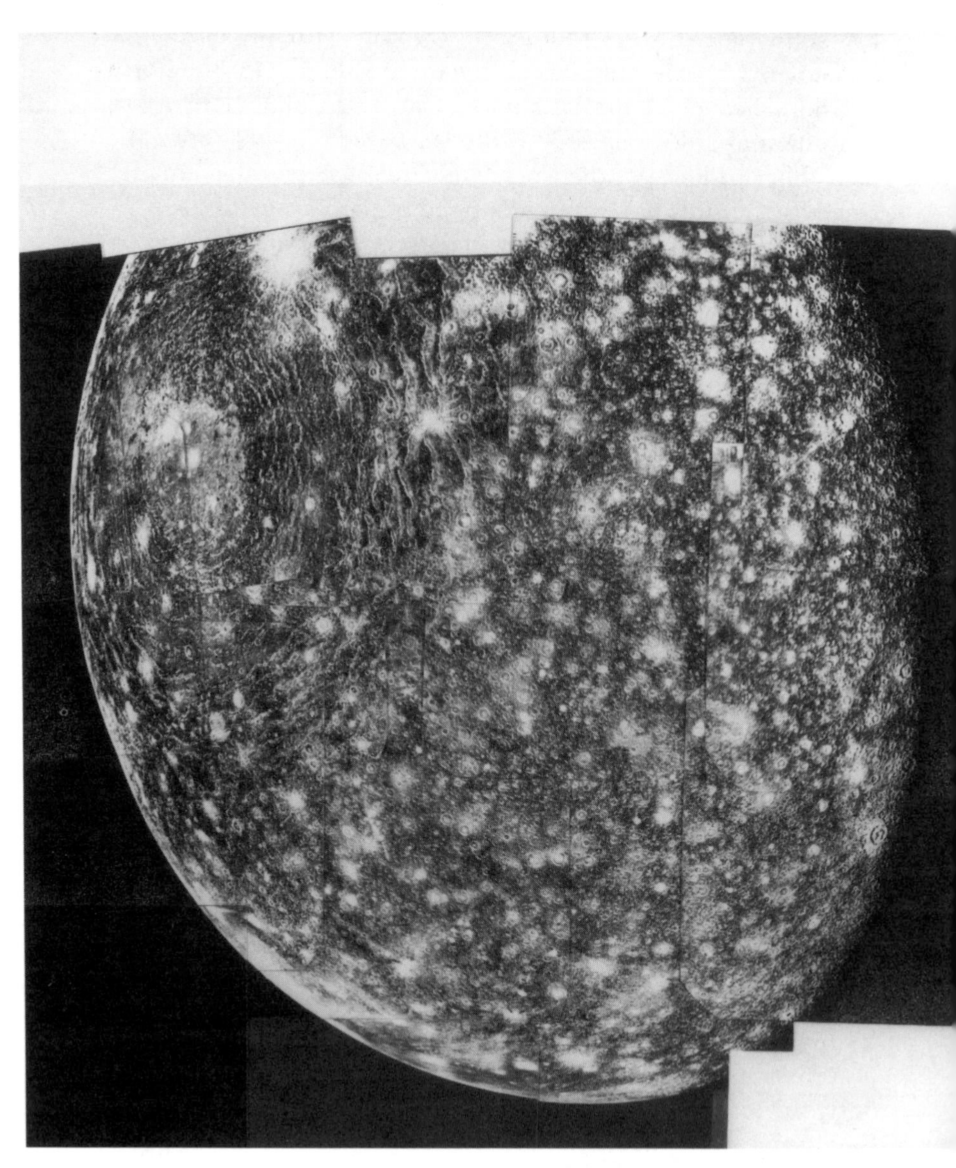

Figure 23. Ringed Structures on Callisto. The large ringed structure seen on the limb is Valhalla. Heavy cratering is visible on the surface. (JPL, NASA)

It is possible that a detailed study of the melting in the central region of Valhalla and the manner in which the material flowed back into the center can set limits that will allow investigators to estimate the thickness of the lithosphere at the time of impact.

Valhalla, at 55° longitude, is located on the leading hemisphere. Just how this large structure contributes to the reduced reflectivity of this hemisphere when it is observed at earth-based resolution is not obvious. However, the crater density and intermediate scale roughness tends to be less in the Valhalla impact region than it is over a major part of the satellite. A mosaic of narrow-angle frames from Voyager 2 reveals a battered surface on the outward-facing hemisphere that is completely saturated with craters.

When the locations of craters that are about 100 km in diameter are noted, it is apparent that no more craters of this size can form without covering others. This would indicate that infall of particles that are large enough to create craters of this size has been frequent enough in the recent past that any evidence of

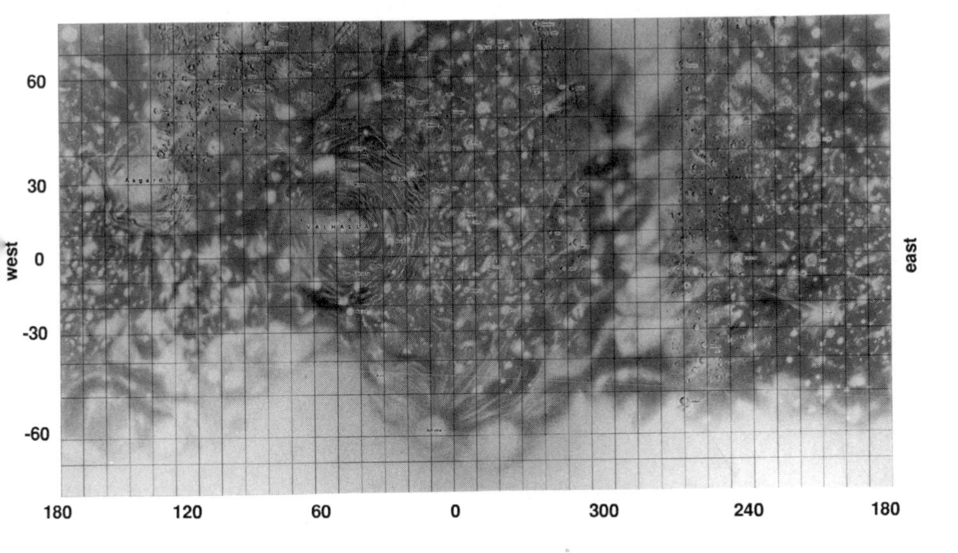

Figure 24. Shaded Relief Map of Callisto. Voyager 1 and 2 data were used to make this map. Longitude 0° is the subplanet point and 90° longitude is at the center of the leading hemisphere. (JPL, NASA)

large earlier impacts on this hemisphere has been obliterated and that the resulting terrain is a ragged landscape of overlapping steep slopes. If there is pulverized rock and ice present, the rough surface would tend to scatter light in all directions; and even if the original material is dark, the powder can be quite reflective. A smooth flow, on the other hand, will tend to reflect the light strongly in a direction away from the incoming beam. This would cause the area to appear dark and may be an explanation of the darkening of the leading face of Callisto.

The Features of Ganymede

As the Voyager 1 spacecraft passed Jupiter and approached to within 115,000 km of Ganymede on March 5, 1979, the satellite showed both dark and light irregular patches that could not be reconciled with the idea that the surface was paved with a dark material and spotted with patches of white ice that surrounded impact craters. During the four hours before closest approach to the satellite, imaging frames showed complex patterns of varying reflectivity and crater density. Detailed inspection revealed that the darkest regions were the most heavily cratered and that the lighter regions displayed grooved structures that cut through the darker material. The general patterns indicated that cracks, or fissures, had developed in the older, darker crust and new, more fluid material had welled up from below, filling the fault zones and flowing and cooling under the local stresses.

Voyager 2 encountered Ganymede before the spacecraft's nearest approach to Jupiter on July 9, 1979, and passed within 62,000 km of it. The resolution was a factor of two better than for Voyager 1, and the synchronously rotating satellite revealed a different, outward-facing side to the spacecraft. A large dark region was observed. High-resolution images revealed that it was heavily cratered, indicating ancient terrain. Along with the craters, this dark region showed a pattern of ridges, or faults, that could have been a segment of an impact, or bull's-eye, pattern similar but larger than Callisto's Valhalla. However, the curvature of the concentric rings is so slight that it is not clear whether this structure is due to a large impact or whether the faults follow great circle routes around the satellite and are a part of a global expansion. This ancient region has been named Galileo Regio, or the Galileo region. Superimposed on the ridged pattern were many small circular craters with central structures. Also present were larger circular structures that showed little evidence of vertical relief even though they appeared to be ancient impact craters.

Voyagers 1 and 2 mapped about one-half the surface of Ganymede. Images obtained at closest approach to the satellite re-

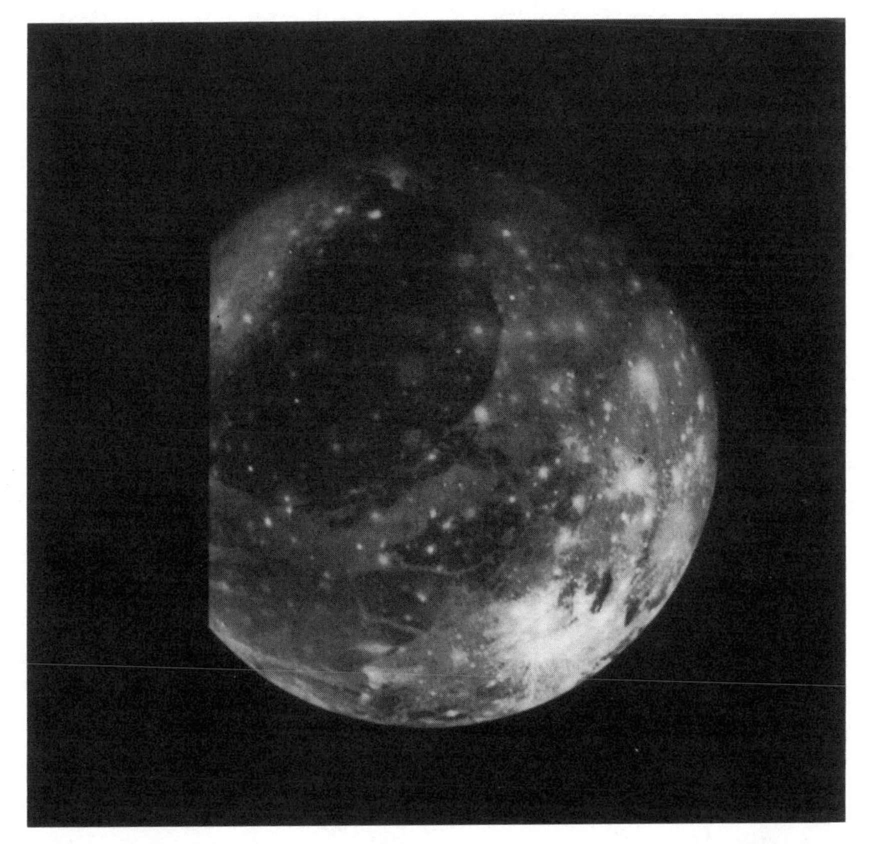

Figure 25. Ganymede. Galileo Regio, the dark area, is surrounded by lighter grooved terrain. Fresh impacts are visible in the lower left and palimpsests are seen in the lower half of Galileo Regio. (JPL, NASA)

solved features with a scale of 500 m per pixel, revealing a surface that had undergone extensive changes over a long time interval. It was readily apparent that the dark polygonal regions were more heavily cratered than the lighter bands of the intervening matrix (see plate 6*a* for Voyager 1 view).

Crater counting, or careful cataloging of the number and sizes of craters in a given area—similar to that done for our moon and Mercury—can provide a framework for establishing the sequence of events that have caused large-scale surface changes. The determination of the actual ages, however, is far more uncertain. On Ganymede the higher density of cratering leads to the assumption that the dark regions are ancient crustal remnants that have been on the surface for the longest period of time. Galileo Regio, located between 90° and 180° longitude and extending over about half the northern part of the leading hemisphere, is typical of the dark heavily cratered terrain. The underlying furrows, with a vertical relief of 200 m from peak to trough, are older than most of the other features in the region and are overlaid by a wide range of types and sizes of craters. Ten kilometers wide and spaced about 50 km apart, individual furrows can be traced for hundreds of kilometers. Other segments of the ancient cratered terrain, located between 42° north and 10° south latitude and 180° to 225° longitude in Marius Regio (named after Galileo's rival), show furrowed patterns that are somewhat narrower and appear to have been displaced and rotated relative to the structures in Galileo Regio. Still, geologists do not have enough information to distinguish whether the formation of these furrowed systems was due to a Valhalla-like impact or a satellite-wide expansion.

Within the ancient terrain there are interesting trends among the craters. The larger the diameter, the flatter the craters appear. Large impact craters seem to have suffered from the inability of the icy crust to support large-scale structures for long periods of time. The smallest craters often appear to be bowl shaped and similar in proportion to craters of like size on the moon. Small well-formed craters with central features are also present. Those with the central peaks range in size from 5 to 35 km in diameter, while those with pits or smoother crater floors are 15 to 120 km in diameter. The difference in morphology is apparently due to the different manner in which craters of different sizes relax in a deformable crust. Relatively smooth regions, which are detectable only because they are brighter than their surroundings, appear as ghosts of larger craters that have collapsed. These features are called palimpsests, named for a parchment-type paper that can be washed and reused. The term suggests that these craters have been smoothed out and are ready to record new bombardment. The palimpsests are almost always located in the ancient cratered terrain.

The other most common type of terrain on Ganymede, covering about 60 percent of the surface, is less cratered and more reflective and grooved than the ancient terrain. It appears to form when the ancient terrain is ruptured and liquid or slushy, icy material is forced up from below, filling the fractures. The grooved details suggest that the deformable material has experienced extensional forces and some rotational shear as it solidified at the surface. The grooves in these areas are separated by 3 to 10 km; the vertical scale tends to be larger than it is in the ancient terrain but generally less than 500 m. The grooves tend to appear as orderly systems that indicate a series of local events have occurred, leading to twisting and crosscutting of the features. The Latin term *sulcus,* meaning groove, has been chosen to describe these linear features.

Other types of geologic features include impact basins. The largest of these, Gilgamesh, is located at 60° south latitude and 120° longitude. It has a smooth area near the center, about 150 km across, which is surrounded by a blocky, rough terrain with a vertical scale of about 2 km. The structure of this basin does not reveal any concentric faults. Instead it appears to have formed after sufficient cooling time had elapsed to allow a thicker crust to form, and may be a younger feature.

Superimposed on the other surface features are ray craters such as Osiris, located at 39° south latitude and 161° longitude. Because these structures overlay others, investigators assume that they are more recent. The distribution of the rays indicates that they have been ballistically formed from material that was ejected during an impact. The higher reflectivity of the rays is from icy debris ejected from the primary impact site that forms a layer of fresh ice and small secondary craters (see plate 6*b*).

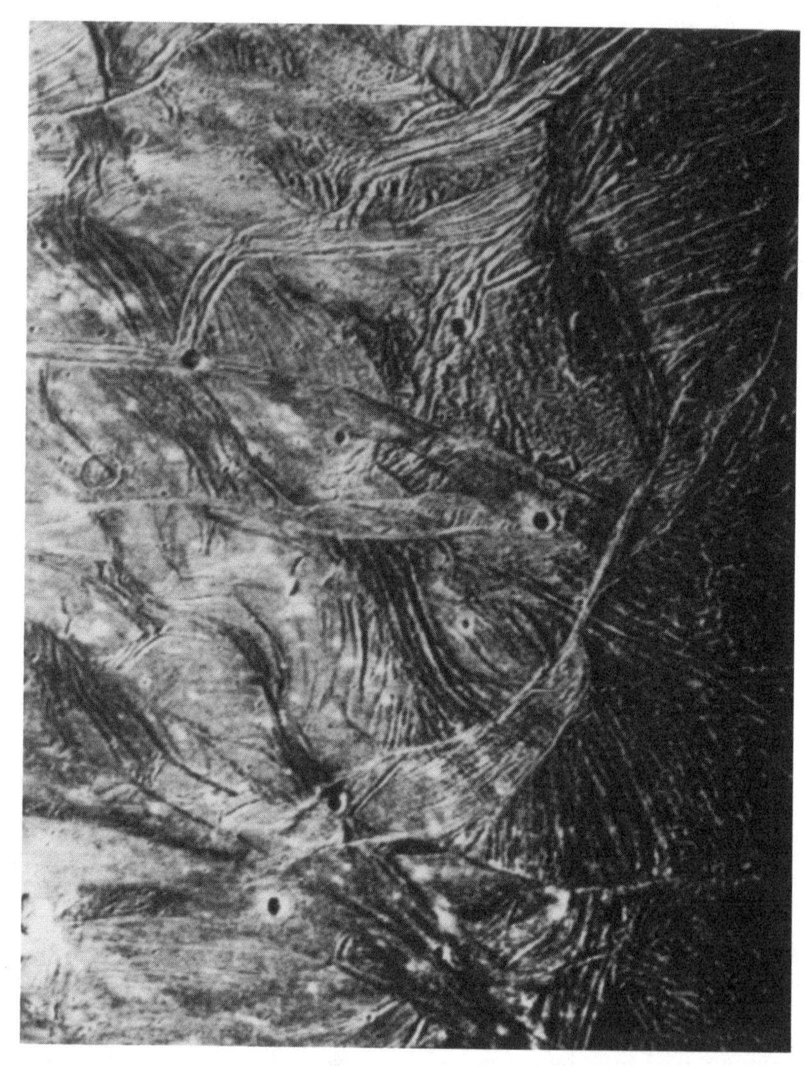

Figure 26. High-Resolution Grooved Terrain. The ridges are spaced 10 to 15 km apart. The pattern of shadows along the terminator, or sunset, line on the right indicates that the vertical scale is about 1 km. (JPL, NASA)

Figure 27. Impact Basin and Ray Crater. Gilgamesh, a large basin with a shattered appearance, is located in the lower right-hand portion of this mosaic. Osiris, a fresh ray crater, is seen in the upper central portion. (JPL, NASA)

The many types of features that are visible indicate that during the early stages of modification of the visible surface, the crust was thinner than at present. Water, or icy "volcanism," grossly modified the surface as the rate of impacts decreased, leaving behind a complicated overlapping record from the various stages of activity and processes that have acted upon the surface.

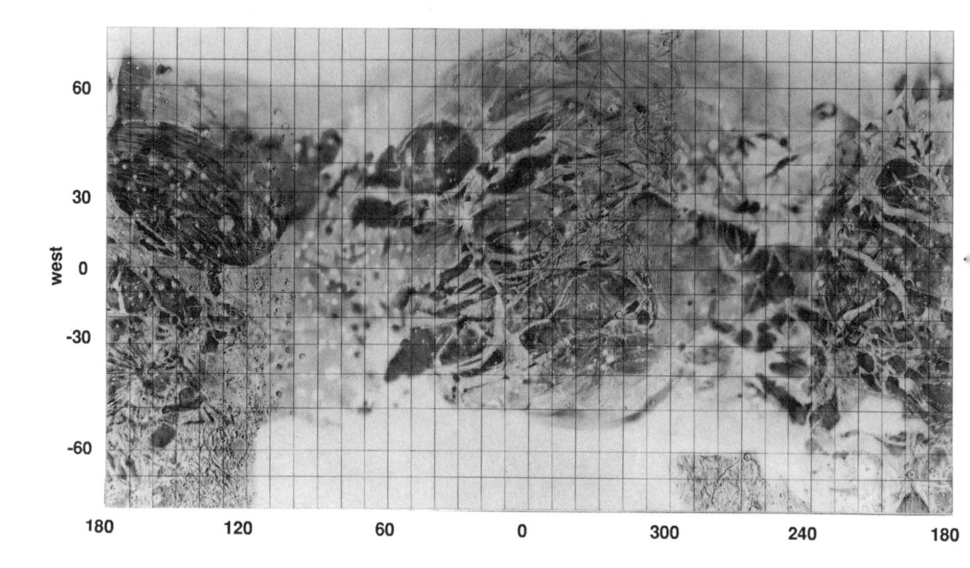

Figure 28. Shaded Relief Map of Ganymede. This map incorporates Voyager 1 and 2 data. (JPL, NASA)

Reasons Why Callisto and Ganymede Differ

Many uncertainties exist concerning the composition of the cores of these icy satellites and the manner in which they were formed. The low average densities of Callisto and Ganymede suggest that water ice must have been a major constituent of the material from which they were formed. Thus the present crust, or lithosphere, may be underlaid by an icy region atop a rocky core. If this is the case, then is it possible to determine the thickness of the lithosphere and the size of the core? It is possible to examine the conditions under which the satellites formed and to ask why Ganymede appears to have undergone more melting than Callisto.

An obvious source of heating is the energy that is deposited during the initial stages of formation of the satellite. When a small body is perturbed into a collisional course with a large body with no atmosphere, such as Callisto or Ganymede, it is accelerated by the mutual gravitational field. The small body reaches a maximum velocity at the instant of impact. The kinetic energy associated with it is equal to one-half its mass times the square of the velocity with which it approaches the large body. At the instant of impact the kinetic energy is deposited into the impact site, causing local heating, melting, and vaporization of material. Thus the extent of alteration of the site is strongly dependent on the impact velocity. However, the manner in which the site adjusts to the impact depends on the strength of the material, the way in which it transfers heat and shock waves, and the depth to which the energy is inserted. Because of such variables, it is not easy to say how much of the heat from the impact will be lost by radiation to space. Scientists do know, however, that if the rate of heating due to this accretion is large enough relative to the cooling rate, temperatures inside the satellite will increase significantly. The material would then become less viscous and the higher density material would sink to form a denser core. Thus a satellite could become layered, or differentiated, and develop a dense core surrounded by an icy mantle.

This same process could have caused a far greater elevation of Jupiter's temperature during the planet's formation. Jupiter would radiate heat, and because this energy would spread out equally in all directions as the planet cooled, heating of Ganymede's environment would be three times more than for Callisto and one-sixth that at Io. Although investigators cannot decisively say, some compositional differences among these satellites may be due to alteration of the medium as they were forming. Nevertheless it is instructive to consider how the satellites have evolved since they were formed.

When the conditions for Callisto and Ganymede are compared, early melting of Ganymede is favored. It is closer to Jupiter than Callisto and more massive, the velocities of collisions would have been higher, and more energy would have been deposited.

The radioactive decay of heavy metals provides another energy source. Some atomic nuclei are not stable and a predictable percentage of the heavy elements will decay within a specified time interval. The ejection of the subatomic particles provides a source of energy to heat the surrounding material. These radioactive isotopes would be more highly concentrated in the rocky, higher density component of the protoplanet nebula. For particles of a given size, collisions within the collapsing nebula would tend to eject less dense particles from the inner regions. The fact that Ganymede is closer to Jupiter than Callisto could allow Ganymede to form with an enriched rocky component. At any rate, the fact that Ganymede has more mass leads to increased heating due to radioactive decay in its interior.

If the initial heat of accretion was enough to cause partial melting, other processes would determine the extent and location of continued melting. The extent to which convective heat transport would occur in the interior would depend on the viscosity of the material. If Ganymede had a larger component of silicate in its mantle than Callisto, the viscosity of the material would be greater. This would retard heat loss through the surface and allow a higher internal temperature to be maintained.

Tidal effects can generate still another source of heating. Tidal heating depends on the eccentricity of the orbit and the distance between the center of Jupiter and the satellite. The fact that Ganymede is closer to Jupiter and is locked into a resonance with Europa and Io suggests that this may be an additional source of heat. However, Ganymede is far enough from the planet that the amount of excess energy that could be generated in its interior relative to that for Callisto is not large. The possibility that initial eccentricities were large cannot be ruled out. Any assumptions that are made along these lines are so uncertain that investigations in this direction cast little light on the understanding of why the evolution of Ganymede and Callisto has been different.

All possible heat sources that have been considered favor more heating of Ganymede. However, the differences between Ganymede and Callisto are small, which suggests that the degree of differentiation—that is, the formation of layers of differing density and composition—may be very sensitive to such influences.

Near the end of the period of growth of a satellite the region around it will be swept out and the number of particles available to deposit collisional energy in the outer region will be reduced. This will allow the surface to cool by radiating heat to its surroundings. The formation of an icy crust contaminated with meteoric material thick enough to absorb local impacts would allow craters to form. The time sequence associated with this is very uncertain; nevertheless, many planetary scientists estimate that the heavy bombardment phase of a satellite's life may have spanned the first half billion years and that the rate of impacts has been much less for the last 4 billion years.

Various scientists have developed computer models to investigate the conditions that may exist in the interior of satellites, and their knowledge of the physics involved in these processes has increased. At the same time analysis of the surface structures on Ganymede has continued in an effort to understand why there has been so much activity on Ganymede relative to Callisto.

At first glance it appears that Ganymede has experienced massive expansion. However, more careful consideration led Steven Squyres, from Cornell University, and E. M. Parmentier, from Brown University, to consider the possibility of a modest expansion which would result in an increase in volume on the order of 5 percent. They argue that the forces due to expansion of the interior would cause a local region to crack and spread apart. Following this, the region between two faults would tend to subside and fluid or slushy material could rise and flood the area. This material would be freer of meteoroidal fragments and would freeze to form a more reflective surface.

This model is compatible with a series of episodic events, and the motion of the flooding material would account for much of the grooved structure. An obvious test of this model involves measuring the thickness of the brighter regions. Are they underlaid with a dark crust that has subsided or are they deep homogeneous flows that would be incompatible with this model? The Voyager images are spectacular but limited in spatial resolution. The color dependence of the reflected light is characteristic only of the surface material. The possibility of obtaining higher resolution data with the Galileo spacecraft and of observing the satellites at longer wavelengths, more sensitive to the material imbedded in the first few meters of the surface, will be discussed in chapter 16.

♃ Chapter 11

Europa, Io, and Amalthea

When the extent of the alteration of Ganymede's surface compared to that of Callisto is considered, and the fact that the tidal forces are a strong function of distance (inversely proportional to the cube of the distance of separation from the planet), it is not surprising that Europa and Io have been grossly altered and that the four innermost satellites appear to be fragmentary remnants of larger bodies.

The diameters of Europa and Io are 3130 and 3640 km, considerably less than those of Ganymede and Callisto. Their masses are, respectively, 32 and 60 percent of Ganymede's mass. Logically, these less massive bodies should not be as dense as their siblings. Surprisingly, even though Europa has a volume equal to only 21 percent of Ganymede's, its average density is 3.0 gm/cc (Ganymede and Callisto have densities that are less than 2 gm/cc). Io is even more dense than Europa, with an average density of 3.5 gm/cc. The size and density of these two bodies indicate that the bulk characteristics should be similar to our moon. Despite this, they are highly reflective, with the reflectivity of Europa equal to 64 percent of the incident sunlight and that of Io, 62 percent. When compared to the moon's 7 percent reflectivity, these are indeed bright surfaces, about as bright as clean snow.

Even though the ground-based infrared spectra of Europa showed strong absorption bands due to water ice, none was seen in the spectra of Io's surface. Other spectroscopic measurements showed the surface of Io to be brighter in yellow-orange light than Europa's and a better absorber at shorter wavelengths. Although these data indicated that the two surfaces were quite different, early observers had ascertained that the leading hemispheres of both satellites were brighter than the trailing sides. This seemed to indicate that as they revolved about Jupiter, their surfaces had been altered by the impacts of small particles. With this early information in mind, the targeting team attempted to maximize surface coverage, obtain as much high-resolution imaging as possible while still searching for unknown satellites, and to characterize the region inside the orbit of Io.

The two latter goals were prompted by a desire to understand the satellites as a system. Scattered light in the earth's atmosphere had severely hampered the study of the inner region, and although Amalthea was discovered in 1892, three new inner satellites were located in the Voyager 1 images. These three satellites were so small that they reflected only about 5 to 10 percent as much light as Amalthea. The fact that the orbits of Andrastea (J XIV), Thebe (J XV), and Metis (J XVI) were unknown before the Voyager 1 encounters precluded their inclusion in the baseline planning, and the high demands on Voyager 2 during the encounter period prevented adjustment of the schedule. As a result the only spatially resolved images of the inner satellites are those of Amalthea, leaving it, the largest of the four, to represent the inner satellites within the Voyager data set. However, comparisons of the relative brightness of the starlike images of the other satellites in several colored filters indicate that the low reflection and deep-red color of Amalthea is characteristic of the group. Despite the limited information about these small satellites, the efforts of the team were highly successful, yielding enough data to allow us to look at the similarities and differences of the inner satellites.

The Surface of Europa

When the decision was made concerning the trajectories of the two spacecraft through the jovian system, a close pass by Europa with the Voyager 1 spacecraft was excluded. The closest approach of that spacecraft to Europa was 734,000 km, and it occurred after nearest encounter with Jupiter. This remote pass limited even the high-resolution narrow-angle camera to a minimum size of detectable features equal to about 14 km. Voyager 1 data did indicate, however, that most of the surface was highly reflective and showed no large-scale regional differences in reflectivity similar to those on Ganymede. Instead, the best Voyager images revealed what appeared to be a network of faults, intersecting each other and extending over most of the surface of the satellite.

The Voyager 2 spacecraft encountered Europa before closest approach to Jupiter and passed within 206,000 km of the satellite. These new, higher resolution images revealed that the icy crust was indeed covered with a network of low ridges and fractures. When the Voyager 2 observations were combined with those of Voyager 1, the resulting data set had the worst resolution of any of the four satellites. Because there were only two encounters and the trajectories had to be chosen well in advance, one of the satellites had to receive less coverage regardless of how interesting its surface might be. Europa was the victim, with only about 40 percent of the surface features mapped at better resolution than 6 km.

A careful inspection of the detailed structure visible on Europa's surface has revealed fewer than twelve small circular craters that could have been formed by impacts of meteoric material. Even though the small circular features could have been formed by some other process, such as steam volcanoes, the fact remained that a small upper limit to the degree of surface cratering could be assigned. This near absence of craters is a surprising result considering the fact that Jupiter's gravitational field should act to focus debris entering the system from outside. When the crater density on Ganymede is compared with the absence of craters on Europa, it leads to the conclusion that Europa's surface is mobile enough to erase all but the most recent meteoric impacts. In other words, because Europa lies deeper in Jupiter's gravitational well, processes on the surface must be generating a well-erased giant palimpsest. Unlike the parchment palimpsests of archaeological significance, which have revealed lost documents of a previous era, this one appears to be almost completely erased.

Views along the terminator, or sunset, line show that the range in vertical structure associated with the ridges is consider-

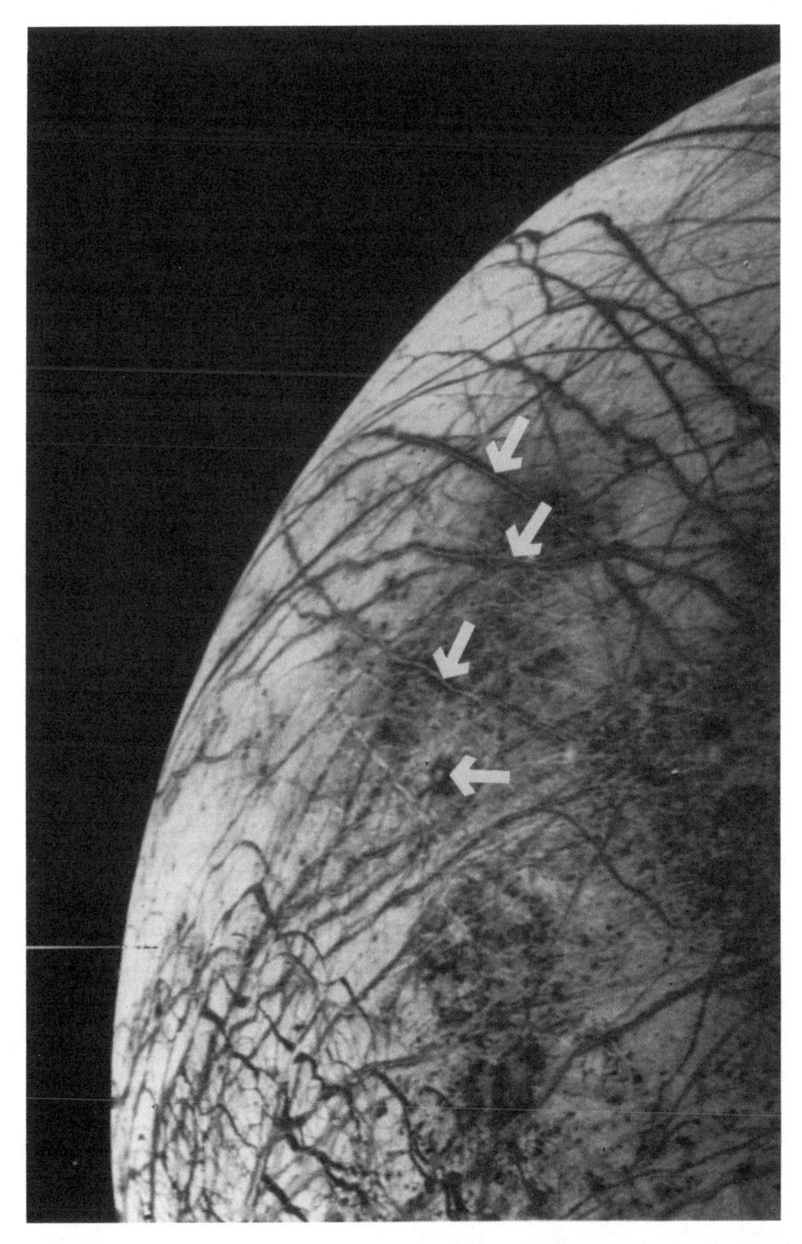

Figure 29. A Crater and Triple Bands. This Voyager 2 image of Europa shows the "triple" bands (diagonal arrows), with a central lighter region bounded on both sides by darker material. A horizontal arrow points to a structure that may be a small impact crater. (JPL, NASA)

ably less than a kilometer. This small vertical scale of the features is surprising because the individual bands are 5 to 10 km wide and extend for thousands of kilometers across the surface. This suggests that either the forces that formed these features had a small vertical component and were directed horizontally or that the vertical structures subside because the surface cannot sustain them.

Based on texture and color, the terrain on Europa can be divided into several types. Plains and mottled terrains are cut through by confusing networks of grooves and ridges. The mottled terrains display two distinct colorations. A brown tint is associated with hummocky angular structures that have a horizontal scale of several kilometers, but as with the ridges, have a low vertical relief, with ranges in elevation limited to hundreds of meters. The gray regions have less horizontal structure and show less distinct mottling.

In addition, the leading hemisphere of the synchronously rotating satellite appears to be bright, while the gray color seems to be concentrated on the trailing hemisphere. Because Jupiter's magnetic field completes a rotation in about ten hours, it will catch up and pass Europa more that eight times per orbit of the satellite. Ions that are trapped in the field will impact the following, or trailing, hemisphere of the satellite. This has apparently caused darkening and graying of the exposed ices on that hemisphere. Thus the variation of the surface brightness of the leading and following hemispheres observed from the earth is due to a darkened following hemisphere, not bright structures on the leading face.

Between 160° and 200° longitude (0° is assigned to the point nearest the planet and 90° is the center of the leading hemisphere), and extending from 20° north to 60° south latitude, there are dark wedged-shaped bands in a region that appears to have undergone considerable faulting in response to local stresses from expansion of the crust. To the north of this region triple bands, dark on both sides with cleaner water ice in the center, trend to the northwest. To the south, the bands trend to the southwest, indicating that the crust has adjusted to a local stress (see plate 5*b*).

The Latin term *linea* is used for the long linear grooves and

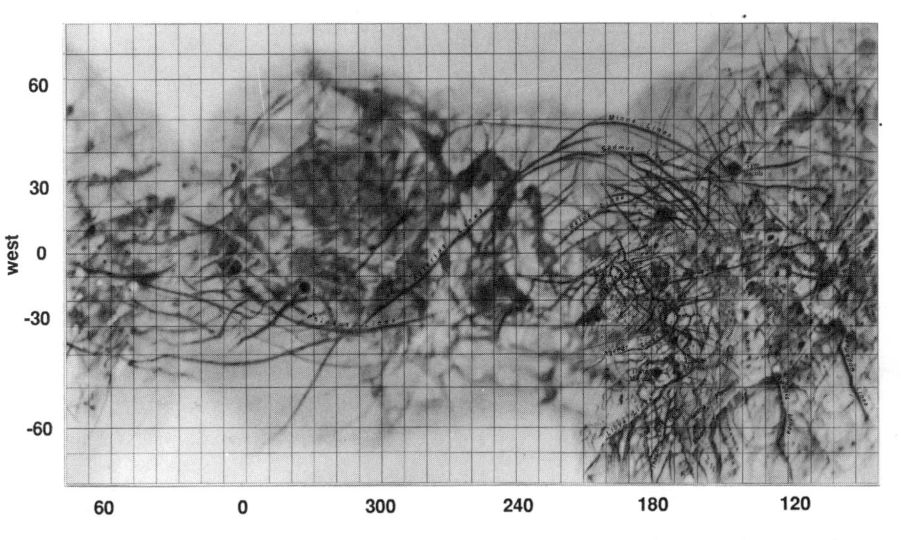

Figure 30. Shaded Relief Map of Europa. Low-resolution Voyager 1 images have been combined with the higher resolution Voyager 2 images. (JPL, NASA)

the term *macula* designates well-defined small brown regions. Thera Macula, at 45° south latitude and 178° longitude, and Thrace Macula, at 44° south latitude and 169° longitude, are irregular brown patches. A third feature, named Tyre Macula, at 34° north latitude and 144° longitude is more circular, and although small, may be the largest impact crater on the satellite. In general, the brown color associated with these areas may be due to material that has been brought up from below or may be residue left behind when ice evaporated from a rock-ice mixture, leaving the brown material exposed.

The average density of Europa is high enough that the bulk of the satellite cannot be water ice. It is probable that the satellite consists of a rocky core overlaid by an icy crust. The many surface fractures are probably caused by tidal stresses, and features like the triple bands are formed by recurring fracturing. These bands may be due to fresh, highly reflective water ice filling in an old fracture zone. The smoother plains appear to be the remnants of an older crust. However, the presence of very few impact craters means that this crust must have been modified or flooded recently enough to have eliminated the evidence of any earlier period of meteor bombardment similar to that recorded on Ganymede and Callisto. Within the framework of the jovian satellite system it seems that dynamic processes could be taking place in Europa's crust. Europa is closer to Jupiter and its period of revolution is shorter than that of the two outer satellites. These conditions would produce stronger and more frequent tidal stresses.

Apparently the energy generated by this never-ceasing tidal flexing has had a profound effect on the thermal history of Europa. In order to maintain an uncratered surface, there must be a balance between the tidally generated heat transported to the surface and that lost to space through radiation processes. The necessary heat loss is accomplished through the icy mobile crust, capable of shifting, flowing, and cracking. The newly exposed surfaces are warmer and radiate away enough energy to allow local cooling to maintain the balance. The fact that Europa still has an icy crust indicates that this process occurs at low enough temperatures to prevent evaporation of the surface ices. Although this balance is delicate, the process appears to have been in operation for a long period of time and to proceed at a rate that is sufficient to wipe out almost all evidence of present and past cratering activity.

Surface Features of Io

The encounter distances from Io for the two spacecraft differed greatly. Voyager 1 approached within 21,000 km of Io after encountering Jupiter. Voyager 2 was limited to passing within 1,130,000 km of Io. Most of the surface area of the satellite was mapped, with Voyager 1 obtaining the high-resolution data and Voyager 2 supplying additional coverage and information concerning short-term changes on the surface. Imaging with a scale of 0.5 to 2 km per pixel was obtained over 50 percent of the surface.

As the spacecraft neared the jovian system, images of Io revealed a peculiar heart-shaped feature located at 20° south latitude and 250° longitude, near the center of the trailing hemisphere. This feature was diffuse and extremely large, appearing to span an area 1000 km in diameter. Later, it was named after the Hawaiian god Pele.

As the spacecraft approached the satellite and the spatial resolution improved, many structures of varying sizes, colors, and textures were visible on the surface. No structure that could be identified as an impact crater could be found. Instead, the multicolored surface displayed features that appeared to have formed from local fluid motion of surface material. Here and there dark holes appeared, showing no similarity to craters on Callisto and Ganymede. During the encounter geologists who were members of the Voyager imaging team were busy selecting images, determining scales of features, and processing data to meet the demands of daily press conferences. Although the geologists were aware that the features they were seeing were formed by volcanic processes, analysis had to wait while the hectic demands of encounter were given first priority. Finally the encounter was over. With the data safely in hand, the geologists went home to sleep and the navigation engineers moved in to prepare Voyager 1 for its journey on to Saturn.

After the spacecraft's closest approach to Io on March 8, as Voyager 1 sped away toward Saturn, the cameras were pointed back at Io to obtain preplanned long-exposure images of its dark side against the background stars. The purpose was to allow the engineers to determine the exact direction of the Voyager 1 spacecraft's path as it left Jupiter. With this knowledge, correction maneuvers could be executed using a minimum amount of fuel. By making the necessary small angular correction near Jupiter, a relatively large miss distance at Saturn could be eliminated. As the navigation engineer, Linda Morabito, processed an image to determine the exact center of Io, she realized that the illuminated crescent had two anomalous features. On the limb of the planet, displayed against the dark sky, was a faint umbrella-shaped structure, and nearer to the center of the disk, at the terminator, or sunset, line, she observed a diffuse bright spot. When the latitude and longitude of these features were compared to earlier pictures, it was determined that they coincided with two of the dark holes, or craters. The unexpected features on Moribito's contrast enhanced image were extensive plumes, indicating that volcanic eruptions were in progress.

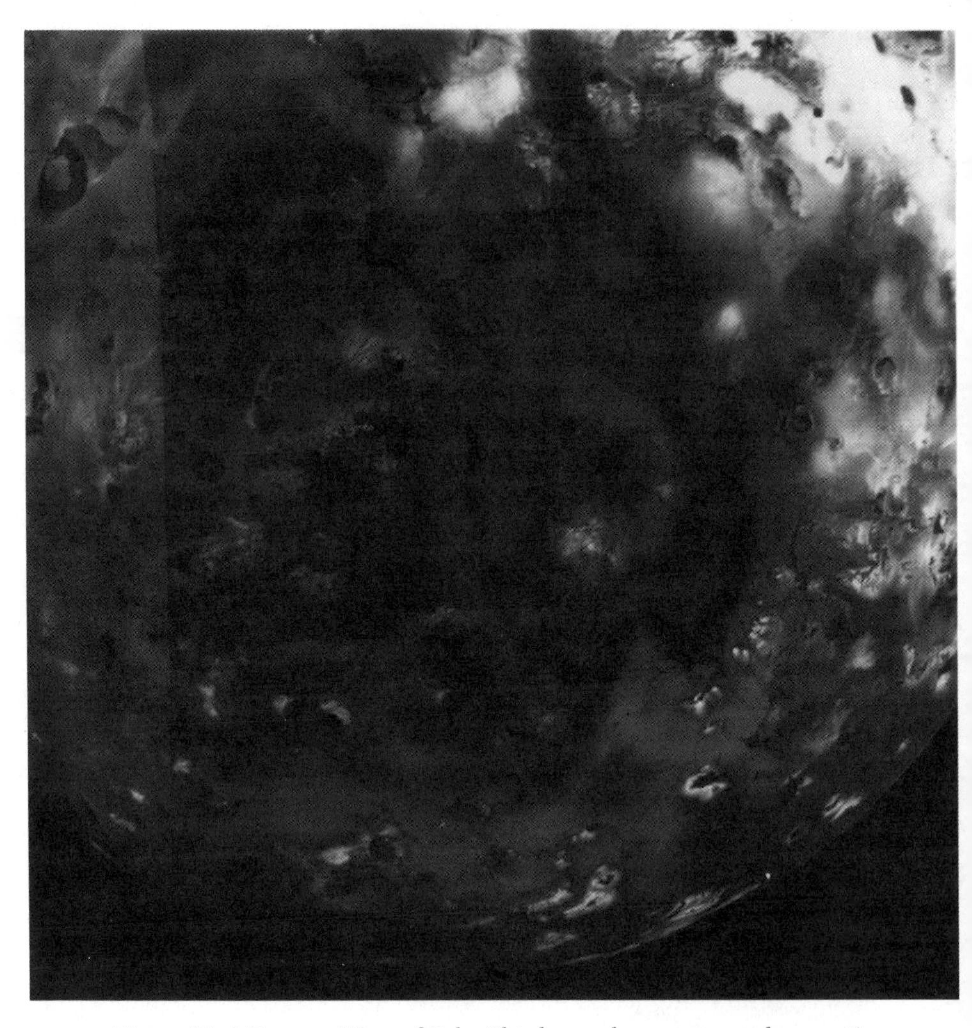

Figure 31. A Face-on View of Pele. This huge plume surrounds an active vent that is located at 11° south latitude and 255° longitude. The heart shape is due to irregularities around the rim of the vent. (JPL, NASA)

The most striking aspect of these eruptions was the fact that they extended 300 km above the surface, which implied that no small forces were driving them. The shape of the plume on the limb of the satellite seemed to show an eruption of hot material emerging from a vent and expanding upward and outward at

Figure 32. Volcanic Activity on Io. This enhanced image shows a plume erupting on the limb and another on the terminator. The low light level on the dark side of the planet is due to scattered light from Jupiter. (JPL, NASA)

high velocities while being decelerated by the downward force of gravity. Judging from the ballistic curves seen in the plume, members of the Voyager imaging team estimated that velocities of 1800 to 3600 km/h were required for the particles to reach the observed altitudes. Flows of this speed would allow the plumes to vent large amounts of energy.

Subsequent inspection of the Voyager 1 images revealed nine erupting volcanoes. Although they appeared to be randomly distributed in longitude, they were concentrated toward the equator. A comparison with Voyager 2 images revealed that eight of the nine were still active four months later, but Pele, the one whose plume led to the discovery of volcanic activity during the Voyager 1 encounter, was inactive; activity near Loki, the Voyager 1 terminator plume, and Prometheus had

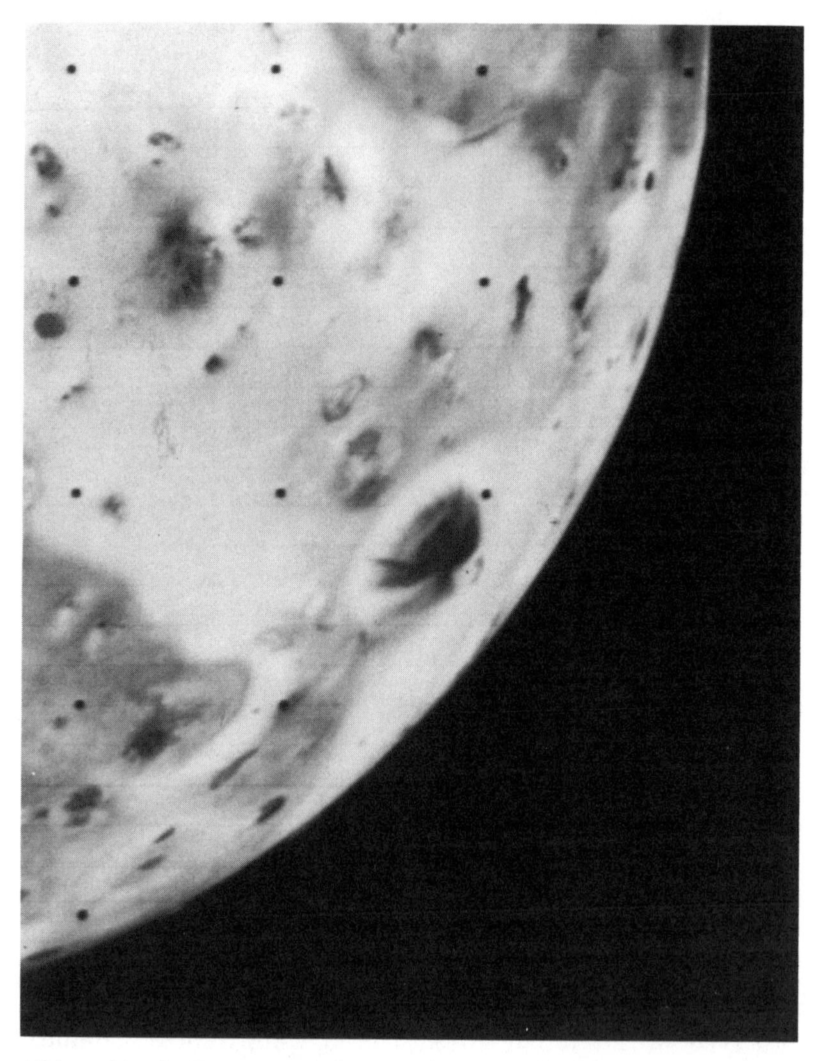

Figure 33 a-b. Two Views of Prometheus. These views of the volcano, located at 30° south latitude and 153° longitude, show it as the line of sight of the spacecraft moved around the planet. (JPL, NASA)

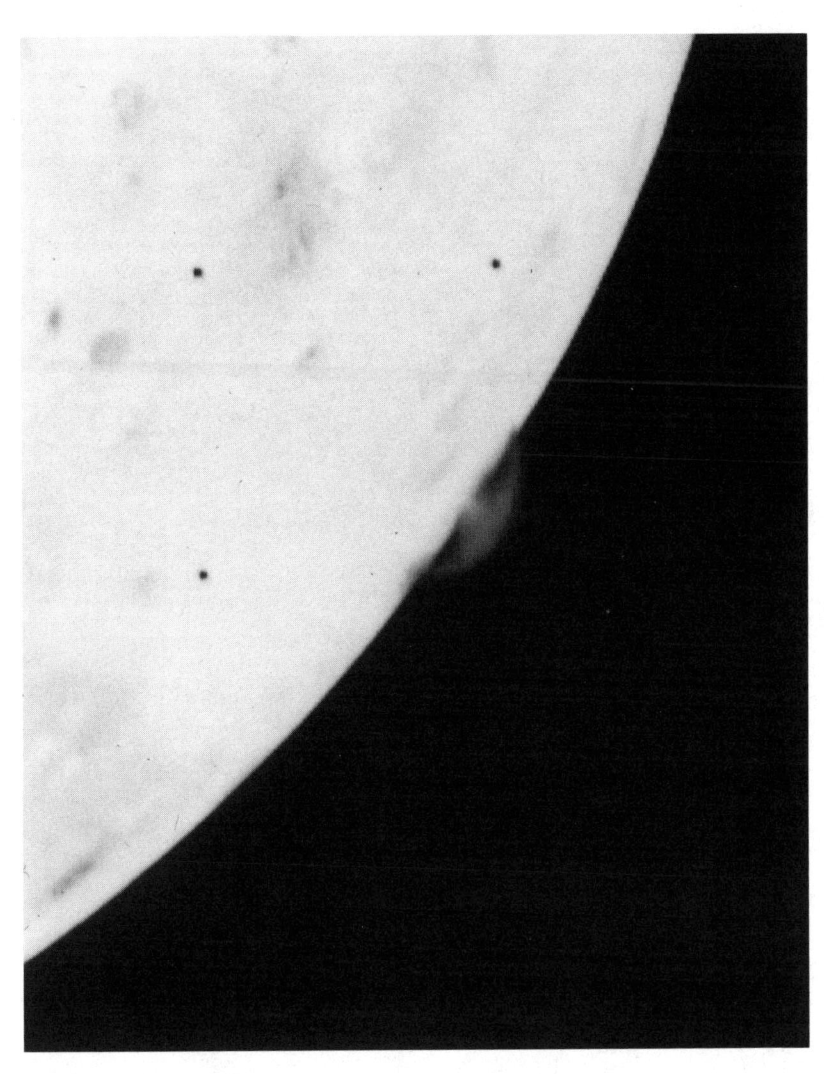

[33b]

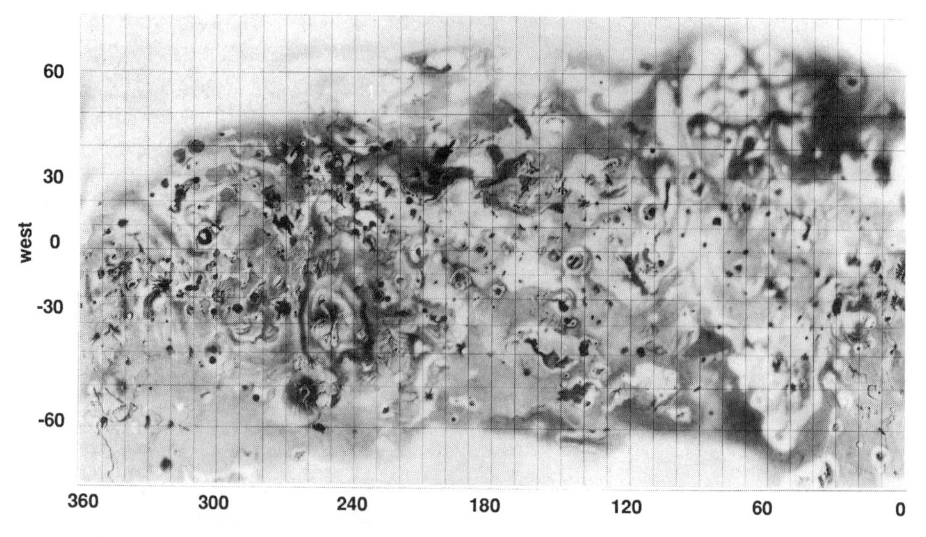

Figure 34. Shaded Relief Map of Io. This map shows the equatorial concentration of the volcanoes. (JPL, NASA)

increased. These results indicate that although the rate that energy is lost through the surface may be constant, venting from an individual site is variable and sporadic. Based on the fact that no impact craters could be found, Torrance Johnson, from the Jet Propulsion Laboratory, and other imaging team members derived a lower limit for the rate of deposition of surface material. They estimated that an average of at least a millimeter of material per year must have been deposited to obliterate all traces of meteor impacts.

The Voyager infrared team reported that its instruments failed to detect even a trace of water ice or steam on or near Io. This contrasts strongly with gases associated with earth-based volcanic eruptions, where the expansion of water to steam contributes to the explosive nature of the eruption. Instead, the infrared team reported spectral features of sulfur dioxide. The Voyager ultraviolet team detected spectral lines of doubly ionized sulfur in the torus associated with Io's orbit (see chapter 15 for an explanation of the torus). These ions could be explained by assuming the sulfur atoms had escaped from the surface of Io and had lost two electrons due to collisions with high-velocity ions that had been ejected from the sun and trapped by Jupiter's magnetic field. Because the number of particles per unit volume is low, the chance that ionized atoms of sulfur would encounter electrons under conditions that would allow them to recombine is very small; hence, once the sulfur atom has lost electrons it will tend to remain ionized.

These two observations that indicated that sulfur was abundant on the surface and was present in the gaseous materials associated with the Ionian eruptions were reinforced by observations by infrared astronomers Fraser Fanale, Robert H. Brown, and Dale Cruikshank, from the University of Hawaii. They detected absorption features in light reflected from the disk that were similar to those caused by low-temperature sulfur dioxide frost in the laboratory. Because sulfur-dioxide frost is white, these observations offered evidence that the surface of the white regions on Io were covered with this material.

The Characteristics of Sulfur Volcanism

In order to address the question of why sulfur dioxide may play an important role in Ionian volcanism, it is helpful to consider conditions in a more familiar domain, the earth's crust. Oxygen and silicon make up about 74 percent of the earth's crust, forming minerals such as quartz or silicon dioxide and other more complex silicates by combining with aluminum, iron, calcium, sodium, potassium, and magnesium—elements that make up another 24 percent of the crust. The fraction of sulfur in the earth's crust is small. When the earth is compared to the sun, it appears that the earth's crust is deficient in sulfur. However, a study of mineral formation reveals that sulfur tends to combine with heavy metals and form sulfides that are dense enough to migrate toward the core of the earth. Therefore, a comparison of Io with the earth does not raise a question concerning an abundance of sulfur in the crust of Io, but about the processes that have allowed it to combine with oxygen instead of heavy metals. This and the importance of silicates in transporting heat to the

Figure 35a-b. Maasaw and Ra Pateras. These two craters demonstrate the nature of the flows from the vents. The scale of the pictures is such that the distance across the bottom image is four times that of the top; thus the size of the caldera are about 25 to 50 km. Maasaw (lat. 40°S, long. 341°W) has a small outflow while Ra (lat. 8°S, long. 325°W) has inundated a region 600 km in diameter. (JPL, NASA)

surface of the satellite are just two of the many problems in chemistry and physics that abound in this alien environment.

When addressing the question of why the volcanoes of Io vent no water vapor, it is important to consider the role that ultraviolet radiation from the sun has played. A photon of ultraviolet light has more than enough energy to break the chemical bonds of simple molecules such as water and sulfur dioxide. The excess energy remaining after the bond is broken is converted into kinetic energy, allowing the two fragments to move away from the scene of the collision. They then collide with other molecules and atoms. As collisions continue to occur, there is a tendency for each type of particle to acquire as much energy as any other type. This is described as equipartition of energy. Because kinetic energy is a function of both the mass and velocity of the particle, the low mass particles have higher average speeds than the heavy particles and have a greater probability of escaping from the gravitational field of the satellite. Because hydrogen and oxygen have atomic masses of 1 and 16 mass units, respectively, it is more probable that hydrogen will escape, leaving the oxygen behind. Oxygen is quite chemically reactive, and as it circulates down to the surface it will combine with the crust, forming oxygen-rich compounds. When sulfur dioxide is decomposed due to interaction with solar ultraviolet light, there is a greater tendency that the oxygen will be lost than the sulfur because the atomic weight of sulfur is twice that of oxygen.

This argument can be used to understand the composition of the crust of Io. Tidal torquing or squeezing and twisting of Io, caused by perturbations on Io's orbit by the other Galilean satellites, has produced stresses that have generated internal heating since the system formed. If heating of the lower crust causes some substance within the crust to change from a solid or liquid into a gas, venting can occur. If a large pressure is built up before

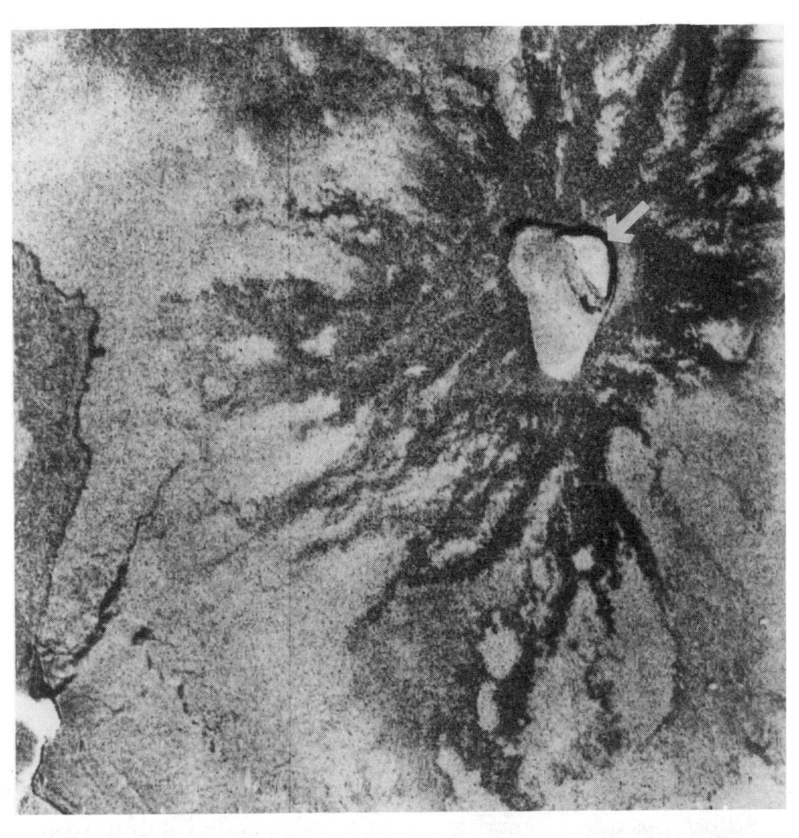

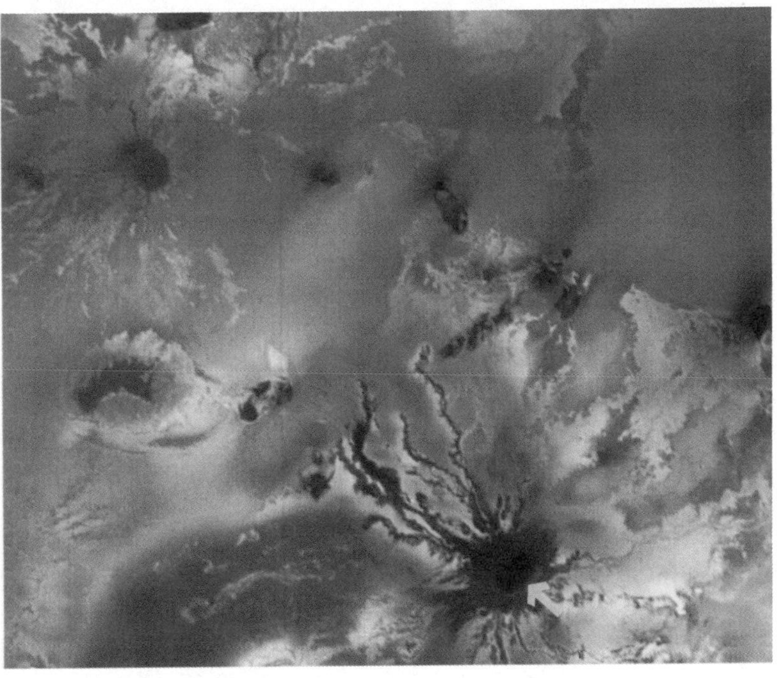

the gas can escape, the rising eruption can spew the material high into the air and create a situation where gas and particles quickly spread out into a large diffuse volume. Then the large surface area of the small particles would quickly radiate heat to space. This is a highly efficient cooling mechanism, but it can work only if there is a source of gas. If there were no available gas, Io would have come to a different equilibrium.

Spectroscopic observations show no evidence of water and indicate that sulfur dioxide is the gas currently shouldering the task of heat transport. If water were present in Io's crust, it would readily form steam and be the preferred driver. The fact that there is no water vapor in the plumes indicates that the crust is desiccated, and thus that the original water was carried upward by earlier venting into a region where it was vulnerable to destruction by solar ultraviolet light. As the hydrogen escaped and the oxygen recombined with the crust, the efficiency with which Io could vent the tidally generated heat would have decreased and the average temperature of the body would have increased until the chemical equilibrium shifted to produce another gas that could act as the driver for the cooling process.

Sulfur dioxide is such a gas. The dissociated sulfur would have to compete with all other atoms to recombine with the oxygen. If this competition led to a surplus of sulfur atoms, they would combine with each other, forming molecules with different numbers of sulfur atoms, or allotropes of sulfur. The number of sulfur atoms in the chain depends of the local temperature. At room temperature sulfur forms a pale-yellow solid, but as it is heated it changes color and becomes a liquid. The color and the viscosity of the liquid change as the temperature increases. Sulfur melts at about 130°C, and the liquid becomes more orange until, near 160°C, it turns to a clear pinkish liquid. At 190°C it appears red, and it becomes more viscous as the temperature increases. When the temperature reaches about 230°C, the molecular structure changes and the sulfur is black and tarry. Near 330°C the material begins to become less tacky, and by 380°C it is a dark fluid. If heating continues, the sulfur becomes a gas. However, the minimum temperature required for this phase change is dependent on the local pressure. Note that the complex behavior of sulfur molecules results in the production of a range of colors, two temperature regimes where fluid flows could occur, 160°–190°C and above 380°C, as well as a gas phase that could drive surface venting.

Although silicate volcanism may be equally important, many of the surface features of Io can be explained on the basis of sulfur and sulfur dioxide venting. A. S. McEwen and Lawrence Soderblom, from the U.S. Geological Survey, reported two different behaviors associated with active venting: explosive, highly variable venting, such as that observed at Pele, and a more gentle, steady venting associated with fluid flows. They proposed that the explosive venting could be driven by sulfur dioxide. If the vent forms a plug that allows the gas to build up until the plug is ejected, the plume-type venting with long-term variable activity could occur. The sulfur dioxide expelled from this venting would condense into solid particles and snow back onto the surface as the clean white frost described earlier. A second, less violent type of venting that is driven by a phase change of sulfur may also occur. It would be more regular and capable of creating sulfur flows and forming detailed structure similar to that seen near Ra Patera. The multiple coloration of sulfur as a function of temperature is in general agreement with the surface coloration of Io, where the hot caldera, or patera, appears black and regions far from the vents are orange and yellow.

The Voyager infrared instrument had a field of view that allowed the team to detect excess heat radiating from several of the patera. The temperature of the Pele vent and the Loki lava lake were about 150°C hotter than their surroundings. Even though these data support the explanation that the Ionian surface coloration is caused by different forms of sulfur, it is not clear what role impurities, such as silicates, play in determining the reflection properties of the surface deposits (see plate 7*a*).

A second question that has been raised concerning silicate volcanism is the role that silicate plays in transporting the internally generated heat to the surface and in maintaining the venting system. The extent to which sulfur and silicate contribute is not easy to determine. Although sulfur-containing gases assail the human nose near any volcanic venting on earth, sulfur flows are extremely rare. In addition, the characteristics of the fluid material vary greatly when sulfur is contaminated with silicate impurities.

One fact that strongly argues for silicate volcanism is the vertical extent of some of the structures on the surface of Io. Measurement of the length of shadows cast by structures on the surface, combined with the known sun-angle, reveals that there are mountains in the polar regions with horizontal dimensions of about 200 km and heights of 5 to 10 km and that some of the steep-walled calderas nearer to the equator are 1 to 2 km deep. Michael Carr, from the U.S. Geological Survey, argues that silicates must be present in the surface layers to provide material that is resistant to flowing under pressure. If sulfur were layered to these heights, he contends, it could not support its own weight and would flow and spread. He points out that the observations of a sulfur-rich surface layer pertain only to the visible surface and offer little information concerning the depth to which this enriched layer extends. Although considerable analysis has been done, many questions concerning the nature of Io's crust are still unanswered, and not surprisingly, there is evidence that the processes that sculpt the surface of Io are complex.

The Interiors of Europa and Io

More than one description of the interior of Europa has been proposed. A comparison of a model proposed by Steven Squyres, which includes an ocean of liquid water under the crust, with one proposed by G. A. Ransford, which binds most of the water into hydrated minerals, illustrates the range of models currently under consideration. If the difference in the densities of Io and Europa is due to dehydration of Io, Europa could be composed of as much as 6 percent water. If the satellite were completely differentiated, this extra water could form a surface layer on Europa that is about 100 km thick. A possible explanation of the low surface relief observed in the Voyager data is that the icy surface is mobile and the rate of flow is adequate to erase impact features. This model would require that the surface ice layer must be several kilometers thick in order for the pressure near the bottom to be high enough to cause the material to flow and generate a glacier field. On the other hand, if there is a liquid ocean below a thin brittle crust, any fracture of the crust would expose liquid water. The liquid water would pour out onto the airless surface and readily flood the area. Although it is not at all certain that Europa has a liquid ocean, if it were true, it certainly would have profound implications. Squyres has pointed out that such an environment increases the possibility that a primitive mobile life form might live in the active fissures. Living near the surface, these simple organisms could take advantage of sunlight and the liquid environment would allow mobility, control of local temperatures, and access to nutrients. Whether this is at all possible is extremely speculative; still, it tickles our imagination.

Models of the interior of Io also vary widely. Shortly before the Voyager 1 encounter with Jupiter, Stanton Peale, from the University of California at Santa Barbara, and Patrick Cassen and Ray Reynolds, from Ames Research Center, published the results of their computer modeling of the interior of Io in the journal *Science*. Their models, which dealt with heating due to tidal effects, predicted that the core of Io would be molten. Following the Voyager 1 discovery of the high level of volcanic activity, Bradford Smith and several of his teammates proposed a model for Io's interior that consisted of a thin brittle crust, or lithosphere, underlaid by a molten sulfur layer several kilometers thick. This layer would lie atop a solid silicate crust that would be about 80 km thick and would surround a molten core. This model was attacked by Michael Carr, from the U.S. Geological Survey, and others, who proposed one that included more silicates near the surface.

Ongoing studies of the effect of the tidal lock among Io, Europa, and Ganymede have been continued by Stanton Peale and Richard Greenburg, from the University of Arizona. Questions concerning how much iron is present and in what form—whether in its elemental form or as sulfides or oxides—have been investigated to understand how variations affect the heat flow from the inside to the surface of the satellite. The role that convection plays in moving the heat from the interior to the surface can greatly affect the results. These are only some of the uncertainties that the investigators face. Although considerable work has been done, there is still a great deal of room for controversial exchange and fruitful investigation. The types of information that can be obtained with the future Galileo Mission that may help us to discriminate among the proposed models for these satellites will be discussed in chapter 16.

Observations of Amalthea

Since Amalthea's circuit of Jupiter takes about 12 hours, the satellite could be expected to pass between the planet and the spacecraft at least once during the interval when the cameras were close enough to resolve surface detail. Voyager 1 passed within 420,000 km of Amalthea, and Voyager 2 came within 560,000 km. This resulted in images with a spatial resolution of about 10 km. At the press conference on March 6, 1979, Bradford Smith, the imaging science team leader, dramatically unveiled the first image of Amalthea, revealing an irregular potato-shaped, or ellipsoidal, body with a cratered surface. It is aligned relative to Jupiter in a manner such that it is in synchronous rotation about its shortest axis. In addition, the axis of rotation is perpendicular to the plane of the orbit, which lies in the same plane as the equator of Jupiter. This results in a configuration with the moon's longest axis radially aligned with the planet. Calculations of time-dependent forces on an ellipsoidal satellite indicate that if a satellite has been in orbit about a large massive planet for an extended period of time, this configuration is the most likely one.

Among the four small inner satellites, Amalthea is the "giant." While the diameters of the other inner satellites are on the order of 40 to 80 km, Amalthea's diametter measures 270 km along the longest axis and 155 km along the shortest axis, and reflects twice as much red as blue light while absorbing 95 percent of the incident sunlight (see plate 4*d*). The satellite's orbital characteristics indicate that it has been in orbit long enough for tidal forces due to Jupiter's gravitational field to have slowed its rotation rate so that it is synchronously locked. However, detailed calculations reveal that the time required for this to happen is short when compared to 4.6 billion years, the current estimate of the age of the solar system. Therefore the synchronous rotation of the satellite cannot be used to discriminate between the possibility that Amalthea is a trapped asteroid or a fragment of original material. The fact that the satellite is an irregular body indicates that it possesses a reasonable amount of internal strength, otherwise it would have been crushed into a spheroidal shape by its own gravity. Its heavily cratered surface, irregular shape, and the presence of the other three small satellites are evidence in favor of the argument that Amalthea is a collisional fragment, left over from the destruction of an older, larger body.

Surface appearance doesn't tell us much about the bulk chemical composition of Amalthea. The low reflectivity of the surface and the dark-red color could be due to sulfur-rich material expelled from Io that has been swept up by Amalthea as the material migrated in through Jupiter's magnetic field. Therefore the color and reflectivity of Amalthea may be related to a thin surface coating and not reveal anything about the nature of its interior. Although questions continue regarding Amalthea's origin and composition, the discovery of the three other smaller satellites and a ring, which we will discuss in the following chapter, are added evidence that Amalthea is a collisional fragment.

When Amalthea, Io, Europa, Ganymede, and Callisto are compared, it is apparent that the inner satellites have been kneaded, outgassed, and bombarded with charged particles trapped in Jupiter's magnetic field. These processes have modified the surfaces to such an extent that there are few obvious clues concerning the initial states of the denser inner satellites. In this sense the jovian system is much like a miniature solar system. Continuing study of the jovian satellite system may yield clues that will assist us to learn more about the manner in which the whole solar system formed.

♃ Chapter 12

Jupiter's Ring

Throughout the past decade the Voyager Mission has obtained fairly detailed observations of the rings of the four planets, Jupiter, Saturn, Uranus, and Neptune. Yet as recently as the mid-1960s astronomers were still unaware that rings were common members of the satellite systems associated with these outer planets. Influenced by observational evidence, investigators had concentrated on explaining why only Saturn had an extensive ring system.

As early as 1659 Christian Huygens (1629–1695) observed the rings of Saturn and described them as a flattened disk about the center of the planet. In 1675 Giovanni Cassini observed that Saturn's rings contained the dark division that was later named for him. He suggested that the rings were made up of small particles in orbit about the planet. William Herschel, who discovered Uranus, disagreed and argued for solid flattened bodies. However, consideration of the range of forces that would be exerted on a solid ring of these dimensions by the gravitational forces exerted by the planet caused other astronomers to favor Cassini's interpretation. Herschel's argument was squelched in 1855 when Cambridge University granted J. C. Maxwell (1831–1879) the Adams Prize for a monograph that presented a mathematical proof that the rings of Saturn could be stable only if they were composed of many small particles.

Although Maxwell's proof was generally accepted, forty years elapsed before observational evidence was obtained. J. E. Keeler (1857–1900) was the first astronomer to obtain proof that the rings of Saturn were composed of small particles in orbit around Saturn. In 1895, using a spectrograph at the Allegheny Observatory, near Pittsburgh, he verified that the particles near the inner edge of the rings orbit the planet at a faster rate than those near the outer edge and that the rate of revolution of the particles at different distances from the planet agrees with the rate that is predicted by Kepler's laws of motion. Therefore observers had evidence that particles too small to be resolved in the telescope were in orbit about Saturn.

Edouard Roche (1820–1883), a French mathematician, considered the problem of growth of satellites out of small particles near a planet. He argued that in order for small particles to grow into larger bodies, the mutual gravitational attraction between two particles when they came into contact with each other had to be greater than the disruptive force that would be proportional to the difference in the gravitational force exerted on the two particles by the mother planet. Because the gravitational force decreases with distance from the center of the planet, there is a critical distance inside which the particles will not stick together, but will be sheared apart by the differential forces exerted by the planet. This distance has become known as Roche's Limit. When this calculation was carried out for Saturn, the observed dimensions of the rings placed them inside Roche's Limit. This indicated that the particles in the rings could not grow and coalesce to form a satellite.

A similar argument can be presented with regard to satellites. It claims that if a large satellite were placed in a close orbit about a planet, the satellite could not withstand the differential gravitational forces exerted by the planet across the diameter of the body and it would break up. When the internal cohesive strength is ignored and only the gravitational field of the satellite and the tidal forces exerted by the planet are considered, this distance is about 2.5 jovian radii for an icy satellite of Jupiter. For denser material such as a nickel-iron alloy, the disruptive distance would be about 1.5 jovian radii. Although these calculations ignore the cohesive strength of the satellite material, they set a rough limit for the distance from the planet where satellite formation could occur.

Astronomers were left then with the interesting and opposing alternates that either Saturn's rings were material that had never condensed into a satellite or they were the remnants of a

Figure 36. Ground-Based View of Saturn Rings. This photograph, obtained on Jan. 19, 1973, represents an astronomer's pre-Voyager concept of the rings. The dark gap in the rings is the Cassini division; the inner bright portion is the B-ring and the outer, less bright portion is the A-ring. (New Mexico State University)

satellite or other solar system body that had become entrapped by the planet. Why then did Jupiter not have rings composed of the small particles or debris?

Ring Models

In an effort to address the question of why Jupiter, Uranus, and Neptune do not have complex Saturn-like ring systems, investigators constructed numerical models. These studies revealed several interesting characteristics of rings. When the investigators constructed a time-dependent model, they established the initial conditions by assuming that there were particles in orbits that were tilted at various angles relative to the rotational equator of the planet. They allowed these orbits to extend over a range of average distances from the planet and to differ in the degree of eccentricity, or elongation.

As they projected the calculations forward in time, they found that the original distribution of orbits changed drastically. Because the particles passed through the equatorial plane twice each time they orbited the planet, they suffered frequent collisions that rapidly damped out the vertical motion and caused them to collapse into a disk. When the distortion of the gravitational field due to the oblateness of the rapidly rotating planet was included, the orbits of the particles tended to precess (their orientation shifted around the planet). The rate of precession was faster for smaller orbits, causing relative motion among the particles. Therefore, when a particle in a large eccentric orbit was near the planet, not only would it be moving rapidly, but it would also continue to encounter a fresh supply of other particles that had precessed into its path.

These collisions would cause a rapid collapse of the ring material into the equatorial region, with a decrease in the eccentricity of the orbits of the particles. The particles would finally be forced into almost circular orbits near the equatorial plane of the planet. The rates of collapse predicted by these calculations were on the order of 10,000 years. When compared with 4.6 billion years, the currently accepted age of the solar system, this is indeed a rapid process. These results indicated that even if a large comet had been destroyed near Jupiter as recently as the beginning of the last ice age on earth, the particles that became entrapped by Jupiter's gravitational field would have already migrated into a ring orbiting in the equatorial plane of the planet.

These models revealed a second problem. They indicated that rings that form in this manner would not be long-lived stable structures. The models showed that collisions among particles would tend to spread the rings and that maximum particle densities would occur in the center of the rings, with the densities decreasing toward both edges. Over a relatively short span of time the rings would spread outward and inward, leading to their demise.

In a general way Saturn's ring structure does agree with this prediction. But sharp edges and gaps, such as Cassini's division, are found among the rings. In 1866 Daniel Kirkwood (1814–1895) called attention to the fact that a particle orbiting Saturn within the Cassini division would have a period equal to about one-half that of Mimas, the nearest satellite to Saturn, and about one-third that of Enceladus, the second nearest satellite. He argued that the satellites could generate well-defined gaps in the rings. Consider particles orbiting Saturn with a period equal to one-half the period of the satellite Mimas. If one of these particles and Mimas were on the same side of Saturn and radially aligned with respect to the planet at a given point in time, the particle would feel a maximum outward perturbing force from Mimas that would shift it slightly in that direction. Exactly one period later, however, when the particle would return to the same point in its orbit, Mimas would have completed only one-half an orbit and would be on the opposite side of Saturn, exerting no outward force. The next time the particle returned to this point, Mimas would be back in position to displace the particle. The process would continue, and on alternate orbits of the particle Mimas would exert a perturbing force that would cause the particle to collide with other particles that were too far from or too near to the planet to resonate with Mimas. Particles would be ground down and their forward motion altered within a region that contained the resonance distance.

When this mechanism, or resonance effect, was proposed as an explanation of the existence of Cassini's division, it was noted that the respective particles in circular orbits near the inner edge of the B-Ring, nearest Saturn, and the outer edge of the A-Ring would orbit the planet in two-thirds and three-halves the time required by Mimas. These resonant interactions could trim the inner and outer extremes of the components of the rings. Although models based on these assumptions did not reproduced all the fine details of the brightness distribution of the rings of Saturn, they could be used to explain the main features.

These studies revealed a second characteristic of the interaction between nearby satellites and small orbiting particles inside the Roche Limit. When a small particle is orbiting faster than the perturbing satellite, as it catches up and passes the satellite in the influence of the planet's gravitational field, it interacts in such a way that it tends to migrate inward. Thus Mimas would aid in trapping the particles inside Roche's Limit, where they could not coalesce into a satellite. None of the major planets other than Saturn was known to have satellites at the proper distance from the mother planet to exert this type of resonating perturbation on particles if they were orbiting within the Roche Limit. By the mid-1900s the question of why Jupiter, Uranus, and Neptune did not have rings appeared to be answered.

The Rings of Uranus

The astronomers' serenity was shattered in 1977 when the rings of Uranus were discovered. Not just rings, but rings with a decidedly different character than those of Saturn. Once again a scientific discovery was made when scientists were looking for something else. In order to plan efficient encounters of the Voyager 2 spacecraft with Uranus and Neptune, these observers were attempting to obtain more precise determinations of the diameters and oblateness of the two planets. As was the case for the Galilean satellites, careful consideration was given to occasions when Uranus and Neptune would occult stars. Because the orbits of the planets were well known, the velocity of the object across the observer's line of sight could be accurately determined. If the combined light from the star and the planet was measured with a high-speed photometer, an observer could determine the times that the loss of light from the star began and ended, and thus easily calculate the distance across the obscuring planet. With the possibility of timing events to an accuracy of one-hundredth of a second, this method would yield more precise dimensions than any previous visual measurement.

An opportunity to measure Uranus occurred in March 1977. Predictions indicated that Uranus would occult a star in the constellation of Libra listed in the Smithsonian Astrophysical Observatory catalogue as SAO 158687. This event would be visible in western Australia and the southern Indian Ocean. A group headed by James Elliot, from the Massachusetts Institute of Technology, obtained observing time on NASA's Kuiper Airborne Observatory. Mounted in a high-flying C-141 aircraft, this telescope would avoid most of the weather problems that plague ground-based sites. Along with several ground-based groups, these observers obtained an unexpected result. Previous to the onset of the occultation and immediately following the reappearance of the star, the starlight was dimmed for short periods of time. The combined observations of Robert Millis, who was observing at Perth, Australia, and of Elliot's group revealed that the planet had a system of several widely spaced narrow rings that were located in its equatorial plane.

Uranus is tipped on its axis in such a way that the line of sight was nearly pole-on during the observation. From the aircraft's point of view, the rings and satellites were revolving around the planet in almost circular paths, like a giant bull's-eye. From the duration of the dips in the light signal, the investigators concluded that the individual rings could be no wider than 10 to 100 km, with gaps between the rings on the order of tens of thousands of kilometers, totally unlike those of Saturn. Subsequent observations by the Voyager 2 spacecraft during the 1986 encounter with Uranus substantiated these results.

The 1977 discovery was made at a time when detailed planning for the Voyager observing sequences of Jupiter was under way, and my colleagues and I were faced with deciding which observations could best answer the question of whether Jupiter had rings. Because the planning of Voyager 1 encounter had already progressed to a stage where the addition of any new observations would require the deletion of other equally desirable observations, it was not possible to plan to verify conclusively the presence or absence of a jovian ring without grossly impacting the whole encounter. However, the discovery of the rings of Uranus, which differed greatly from those of Saturn, had dealt a staggering blow to well-accepted ideas about the structure and stability of planetary rings. A search for jovian rings was needed.

Tobias Owen, a member of the imaging team now at the University of Hawaii, argued that even though the necessary observations would occur during the most intense observing period of the Voyager 1 encounter, some effort must be made to search for

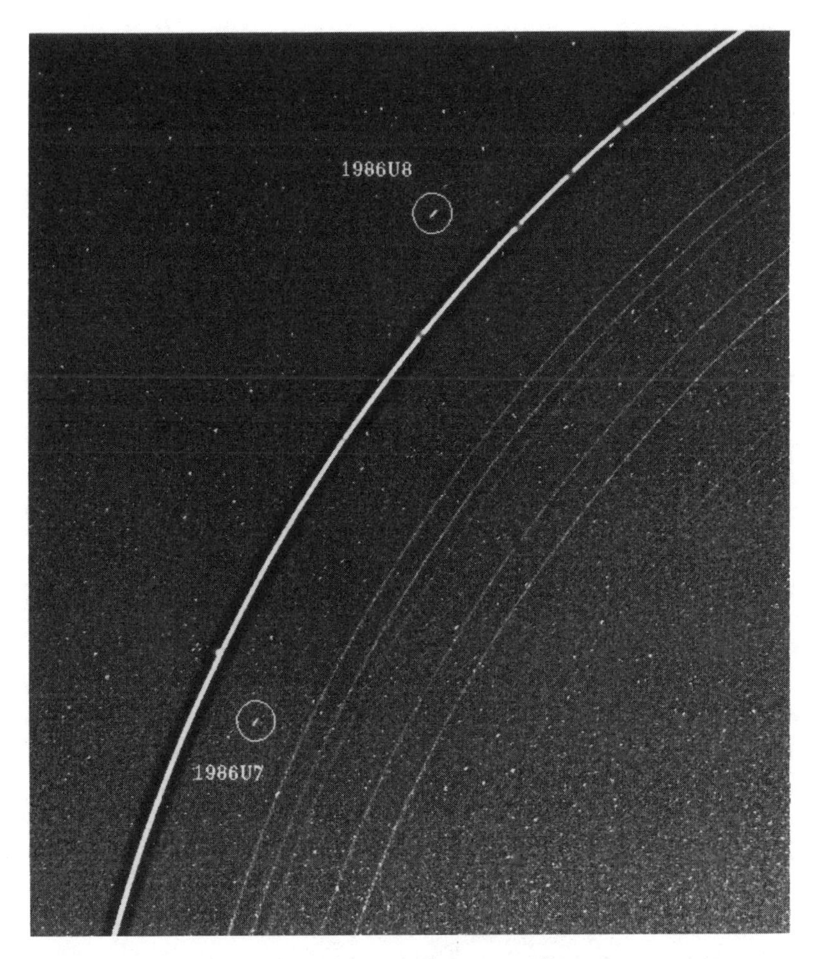

Figure 37. Voyager View of Uranus' Rings. The image processing to bring out the faint rings generated a dark halo about the bright epsilon ring. Two of the small moons are highlighted. (JPL, NASA)

an equatorial ring. If it were found during the first encounter, more observations should be obtained with Voyager 2 to establish its nature. Others argued that a possible ring was less important than detailed information about the surface of the satellites and that if a ring existed, it should have already been detected by earth-based facilities. The latter argument was weakened when ground-based infrared observations indicated that the Uranian rings were very dark and possibly were composed of material similar to that of asteroids. If Jupiter had a ring of this dark material, it would not be easily observed from the earth.

Armed with calculated values for the minimum surface brightness of a jovian ring that could be detected with earth-based telescopes, Owen argued that a long exposure was needed to search for such a ring. Because the numerical models indicated that particles in orbit around an oblate planet must collapse into the equatorial plane, the Voyager imaging team members agreed that the optimal scheme for detecting a very faint ring would be to take a long exposure when the line of sight of the camera was aligned along the equatorial plane of the planet. If the ring were in the equatorial plane, it would be seen edge-on, minimizing the light that would pass unscattered through the rings. This would yield the largest possible fraction of the light being back-scattered into the cameras. Careful arguments indicated that an exposure of at least 11 minutes (in comparison to 0.3 second exposures of the atmosphere of the planet) would be required to obtain a high-resolution image with a narrow-angle camera.

Because there was no way of estimating how large the ring structure might be, the Voyager imaging team decided to use the diameter of the rings of Saturn as a guide and to look in the same general region of the equatorial plane of Jupiter. As a result, a single simultaneous wide-angle and narrow-angle pair of frames of the region was scheduled. The wide-angle frame covered the region out to the orbit of Amalthea, but the field of view of the narrow-angle frame was limited. The team members knew that the wide-angle camera would pick up light closer to the planet and would suffer from scattered light within the optical train of the camera. But because the spacecraft had to be pointed and the exposure time for the narrow-angle frame dominated the sequence, they decided to include a wide-angle frame anyway. They were worried about missing the ring location and hoped to get lucky with a partially useful wide-angle image. More than a year before the arrival of the spacecraft at Jupiter, the sequence to search for a possible ring took its place in the crowded near-encounter observing schedule.

The Ring Discovery

After traveling along a curved trajectory for about 135 billion miles, the Voyager 1 spacecraft approached to within 1.2 million miles of Jupiter on March 4, 1979. After thousands of images had been acquired, the time finally came to point the cameras at the ring target area. The two cameras were centered on the same point, but as expected, the fact that the field of view of the wide-angle camera was seven and a half times wider caused the scattered light to fog the wide-angle image so badly that it was useless. The narrow-angle frame, however, did not suffer the same fate. When it appeared on the monitor before the waiting team members, it contained a strange pattern. The stars in the field of view did not appear as straight trails as they should with an 11-minute exposure. Instead, there were patterns that resembled parts of hairpins. Consternation reigned briefly. Then one of the team realized that the pattern was due to spacecraft motion. Because some spacecraft components, including the magnetometer and the generators, are located on long booms, there is a slight oscillation about the spacecraft's center of mass. The period of oscillation is about 76 seconds, with the spacecraft rocking back and forth as it moves forward in space. No other exposure had been long enough to reveal this pattern, but in this image the hairpin patterns were generated by the combined oscillation and forward motion of the spacecraft during the 11-minute exposure.

A second pattern was superimposed on the hairpins. Six faint narrow bands ran diagonally across the frame. Someone realized that this pattern was consistent with a smeared band of light extending across the field of view. Bedlam reigned. To have hit a ring with such a limited field of view was astounding.

Immediately the question arose, How far was the ring from Jupiter? The team knew that uncertainties in pointing could be as large as a quarter of the field of view of the narrow-angle camera. Where was the camera pointed? Predictions of the trajectory indicated that the line of sight past Jupiter toward the stars was such that the Beehive cluster in the constellation of Cancer would be behind the planet. Several stars from this rich open cluster should lie in the field of view. The team scattered like house mice on Christmas morning to obtain star maps and to determine precisely their line of sight. When Bradford Smith, the imaging science team leader, announced the discovery on March 7, 1979, the members of the waiting press responded in amazement to the sheer good luck of imaging the edge-on ring with a single narrow-angle frame.

Inspection of the Voyager 1 image revealed that the outer edge of the ring was located at about 1.8 planetary radii from the center of the planet and that the radial extent of the ring was 57,000 km. The resolution of the discovery image was such that it was possible to say that the main part of the rings could not extend above or below the equatorial plane by more than 20 km. This image also revealed that the radial extent of the rings of Jupiter was less than that of the rings of Saturn and that the rings scattered only a small fraction of the incident sunlight back into the cameras. There was no evidence of fine structure: imaging of the details within the rings would have required a line of sight that was inclined relative to the equatorial plane of the planet.

There was not sufficient time after the discovery of the rings to alter the observing sequence to allow the Voyager 1 spacecraft to look back after passing by Jupiter to see the rings at a different inclination. The sequencing team did have ample time, however, to program Voyager 2 to target the rings during the July 1979 flyby. The reflectivity of the rings in the Voyager 1 discovery image led the team members to suspect that they were composed of widely spaced small particles in orbit about the planet in a configuration that would allow most of the light to pass on through. Because Jupiter has four large satellites as well as Amalthea and other possibly undiscovered small satellites in orbit near the planet, the particles in the rings could experience variation in gravitational forces and could be subject to resonance perturbations similar to those that Mimas exerts on the Saturnian rings. Collisions of the perturbed particles with nearby less affected particles plus collisions with high-energy ions that are trapped in the magnetic field would produce a steady supply of small particles. These small particles would not evaporate rapidly, and a reasonable population of small particles should exist in the rings.

Anyone who has observed sunlight scattered by morning mist,

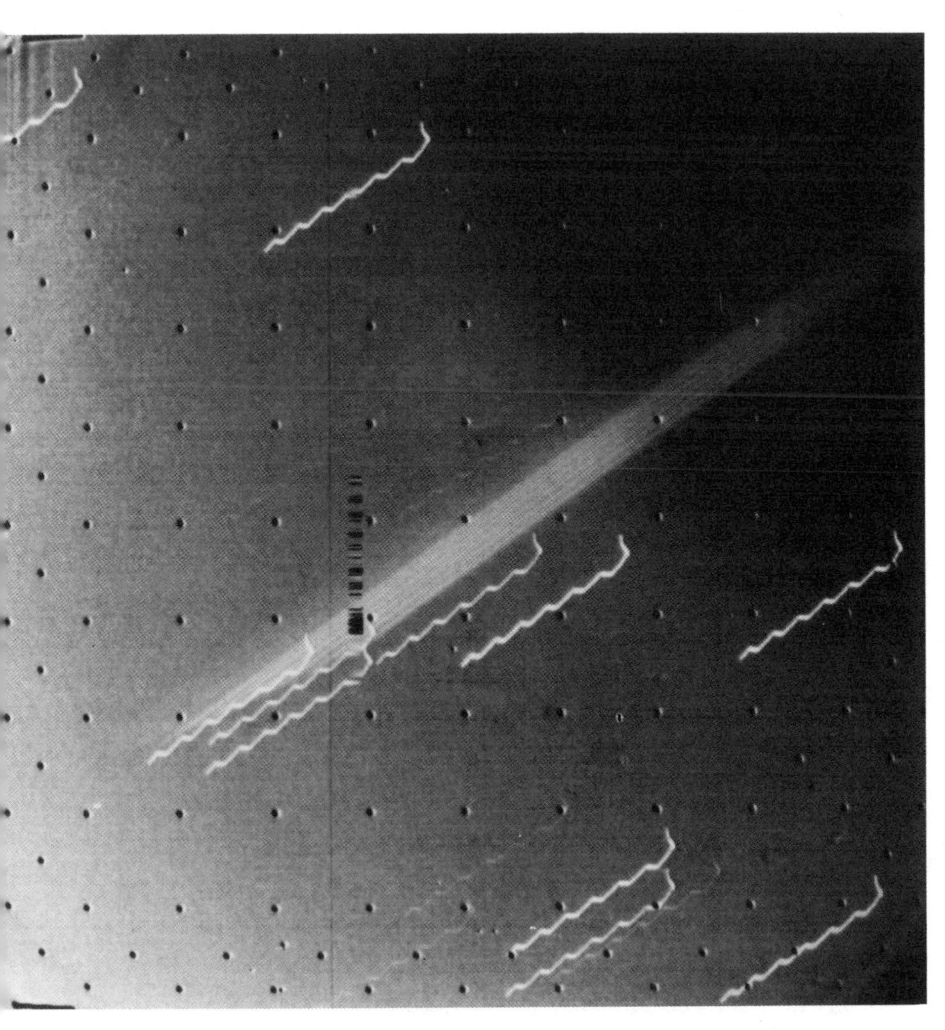

Figure 38. The Rings of Jupiter. This 11.2-minute exposure at a distance of 1,212,000 km was the means of discovering these rings. The hairpin patterns are background stars, smeared by the spacecraft's motion. This motion included a 76-second oscillation that caused the wiggles and a sixfold exposure of the rings, which stretch diagonally across the center of the frame. The black dots were placed on the camera to allow for removal of geometric distortions in the videcon and the shading on the left is due to Jupiter's scattered light. (JPL, NASA)

southwestern dust storms, urban smog, or condensation trails from commercial airlines is aware that the visibility of the small particles varies greatly, depending on the angle between the sun and the observer. Small particles scatter weakly back into the direction from which the sunlight is coming and strongly in the opposite, or forward, direction away from the sun. The imaging team reasoned that if they scheduled the observations after the spacecraft had passed the planet, the forward scattered light would allow greatly reduced exposures. This would allow the team to obtain sharp unsmeared images that would show the radial structure of the rings when the line of sight was at an angle to the equatorial plane. Sequences were designed that commanded the ship to obtain several observations of the rings when the spacecraft was at a range of about 1.5 million km (one-hundredth of an Astronomical Unit) and within the shadow of the planet, so that the sun could not shine into the sensitive instruments on board the spacecraft. The resulting observations revealed that the jovian ring had a sharp outer boundary and shaded gradually in toward the planet, totally unlike the Saturnian rings or the thin, widely spaced Uranian rings. The images showed disappointingly little structure.

Structure of the Ring

Tobias Owen and his co-investigators analyzed the manner with which the brightness of the ring varied. Their results showed that the largest density of the particles was along the outer edge of the ring and that the particle density appeared to decrease gradually inward toward the planet. The combined data from Voyagers 1 and 2 revealed that the rings appeared twenty times brighter in forward scattered light than in the reflected light measured by the spacecraft as they approached the planet. These results verified the earlier speculation that the rings contained a component of small particles that scattered strongly in the forward direction. The investigators concluded that the ratio of the forward and backward scattering indicated that the diameters of the particles were a few microns (a micron is one-millionth of a meter), similar in size to particles that make up

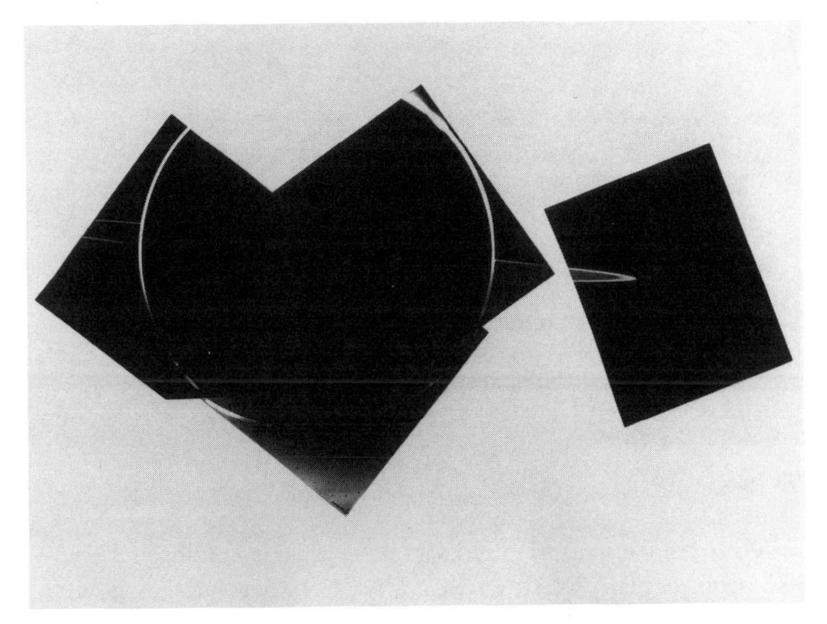

Figure 39. Forward Scattering through the Rings of Jupiter. This mosaic was taken with the wide-angle camera when the spacecraft was 1,500,000 km beyond Jupiter, within the shadow of the planet. The brightness of the rings and halo around the planet are due to efficient scattering by the small particles in the rings and molecules in Jupiter's atmosphere. (JPL, NASA)

the hazes in the earth's atmosphere. The maximum surface brightness of the rings compared to that of the jovian cloud deck is about 1 to 30,000. The color of the reflected light is similar to that from the surface of Amalthea, indicating that even though many small particles have been formed from collisions of larger particles with tiny incoming meteorites or charged particles, the larger surviving particles must have a surface coating of the same dark-red material that forms the surface of Amalthea.

The fact that the inner boundary of the ring is diffuse and material is visible all the way to the cloud deck of the planet indicates that material is spiraling into the planet's atmosphere. A particle in orbit around Jupiter at a distance of about 1.8 planetary radii must have a period of nearly 7 hours, while charged particles that are trapped in the magnetic field are

Figure 40. Structure of the Jovian Ring. These images were obtained 26 hours after closest encounter. Voyager 2 was looking back from an angle of two degrees below the plane of the rings. This view shows a gradient in brightness inside the rings that extends inward toward the cloud tops. (JPL, NASA)

forced to rotate with the field and have a period of about 9 hours 55.5 minutes. This would cause collisions between the ring particles and the charged particles that would reduce the forward orbital motion of some of the ring particles, causing them to spiral into the planet. At the same time this interaction with the charged particles would erode the ring particles through collision, generating a continuous new supply of small particles to replace the ones falling inward.

Additional analysis of the ring images by David Jewitt and G. Edward Danielson, from the California Institute of Technology, revealed that small particles extend about 10,000 km above and below the ring. They interpret the data as evidence that collisions with charged particles cause disruption of the ring. Because the ring particles revolve about the planet in its equatorial plane perpendicular to the axis about which the planet rotates, and because the axis of the magnetic field is tilted relative to the rotational axis of the planet, collisions with the charged particles moving with the magnetic field scatter material out of the ring. These neutral particles will orbit the planet in inclined orbits, passing through the ring two times each orbit, until collisions with other more equatorial particles damp out their north-south motion.

The sharp outer edge of the ring suggested the presence of a moon at the right distance and with a large enough mass to generate a shepherding effect on the ring particles. Among the known satellites, Amalthea was the most likely candidate. Particles near the edge of the ring circle Jupiter about three times each time that Amalthea revolves twice. However Amalthea, in orbit about 2.5 planetary radii from the center of Jupiter, is too small to confine particles and retain the sharp boundary of the ring. A new satellite was needed to explain the observations. Luckily, observations had been planned to allow a careful search of Voyager images for small, faint satellites near the planet. It was this search that revealed the three small satellites in orbit near Amalthea. Metis and Andrastea, with diameters of about 40 km, orbit Jupiter just outside the edge of the rings. The members of the team proposed that the two satellites play an active role in trimming the outer edge of the rings. (The fourth inner satellite, Thebe, at 1.3 radii beyond the ring edge, is too small to contribute effectively to the process.)

After the excitement of the encounter subsided, interpretation of the data was carried out at Cornell University by Mark Showalter and co-workers. They pointed out that it was not necessary for the density of the particles to decrease gradually in toward the planet. If the ring was a torus, or doughnut-shaped, cloud with a concentration of particles along the outer perimeter due to the constraining influence of Metis and Andrastea, the resulting image would look much like the one observed. Along with Joseph Burns, also from Cornell University, and Jeffery Cuzzi and James Pollack, from the Ames Research Center, Showalter constructed models of the rings that were based on observed brightness distributions. These investigators modeled the interaction between a distribution of ring particles and charged particles that were trapped in the magnetic field. They concluded that the rate of the loss of mass from the rings due to particles spiraling into the planet is small and that the total mass of all the particles in the rings is equivalent to the mass of a satellite 70 km in diameter, much more than that contained in Comet Halley. Their model, based on realistic assumptions concerning the composition and size of the ring particles, predicted that less than 20 percent of the total mass of the ring would have been lost during the lifetime of the solar system. These results indicate that one cannot distinguish whether the current thin

Figure 41. Voyager Looks Back on the Rings of the Outer Planets. These views show a lack of radial structure in Jupiter's ring (upper left) and the complex structure of Saturn's rings (upper right). These rings contrast with Uranus' widely spaced thin rings (lower left) and the lumpier rings of Neptune (lower right). Note the increased noise due to weaker signals at Uranus and Neptune. (JPL, NASA)

ring is a residue of original debris that has been exposed to rather gentle satellite perturbations throughout the lifetime of the system or is material that has been acquired more recently. The rings could be composed of a mixture of cometary material, collisional ejecta from meteors colliding with the satellites, and material expelled from the volcanoes of Io.

Within two years after the discovery of Jupiter's rings, groups headed by Robert Millis, from Lowell Observatory in Flagstaff,

Arizona, William Hubbard, at the University of Arizona, and Andre Brahic, from the Observatory of Paris, in Meudon, France, detected what appeared to be incompletely filled ringlike structures encircling Neptune. The 1989 encounter of Voyager 2 with Neptune revealed that this was indeed true, and here was yet another type of ring structure. Although Neptunian rings are narrow and widely spaced, the distribution of the material around the ring is clumpy, not smoothly distributed as in the Uranian rings.

The scientific implications of the Voyager and ground-based observations of the ensemble of ring structures surrounding the four planets are that complex systems of satellites and rings are characteristic of the outer solar system, and close to the planets, the satellites become smaller and the moon-ring interactions become very complex. Current and future analysis and modeling will deal with the evolution of these complex systems. As Jupiter's ring system takes its place among a family of small objects encircling bright planets, scientists have become aware that the dynamic interactions that sculpt the rings are far more complex than they had suspected before they obtained the high-resolution Voyager data. Were all the rings formed billions of years ago or did some of them form out of recently acquired cometary residue? As dynamists study the interaction of the rings and imbedded moonlets, will they learn more about how the planets were formed? Will investigators succeed in using the orbiting Galileo spacecraft to make observations in the late 1990s, and will these data answer the questions about the jovian ring?

♃ Part IV

THE MAGNETOSPHERE

♃ Chapter 13

The Discovery and Investigation of Jupiter's Magnetic Field

Although the life of the average reader will not be influenced by the state of Jupiter's magnetic field, this is decidedly not the case for the sun. How the earth's magnetic field shields us from high-energy particles that have been ejected from magnetic storms on the sun is of considerable interest to us. Interference with radio communications and radiation damage to astronauts working outside the earth's magnetic shield are a part of everyday reality. In Jupiter investigators have a planet whose mass is three hundred times that of the earth but only one-thousandth that of the sun. This rapidly rotating planet provides an intermediate set of conditions for testing their ideas of how magnetic fields are generated and maintained in our own planet and the sun.

Some of the effects of the ways energetic particles modify a system have been illustrated in the last three chapters. There we noted that charged particles had played an important role in modifying the surfaces of the inner satellites and perturbing the small particles in the rings. We now consider the nature of Jupiter's magnetic field and review our attempts to explore it. In this chapter we recall the discovery that Jupiter is a strong source of radio signals and examine the historical evidence that revealed that Jupiter has a strong magnetic field. In chapter 14 we will consider how a magnetic field is generated and maintained within a planet and will describe the externally generated structure, the magnetosphere. Armed with this insight, we will then review the Pioneer and Voyager observations in chapter 15.

Jupiter's magnetic field at the cloud tops is similar to that at the earth's surface. Although it is about twenty times stronger than the earth's field, as on earth, the poles are tipped about 10° relative to the axis of rotation of the planet and the magnetic axis does not pass through the center of the planet. Even though these characteristics do not seem alien, scientists still have far to go to understand the nature of this strong field. Because the magnetic field is generated in the interior of the planet, detailed observations of it can provide indirect information about Jupiter's internal structure. Thus the chance to fly close to Jupiter with the Pioneer and Voyager spacecraft and to measure both the field strength and the particle densities promised to provide a mother lode of information about this complex system.

The basic data that astronomers use to detect the magnetic fields of other planets consist of time-dependent radio signals of varying strength and frequency. These data can be acquired by constructing large tunable earth-based receivers. Spacecraft can obtain similar data during flybys or orbiting missions, but with the added advantage of simultaneously measuring the number of particle impacts experienced by the spacecraft. Instruments for particle detection are designed to measure the direction, mass, and energy of charged particles that are trapped in the magnetic field. With a spacecraft there is an additional advantage: investigators can obtain data at different distances from the planet and at varying orientations relative to its rotational axis.

Over the past century recognition of the mechanisms that generate a global magnetic field and knowledge of the processes involved in producing the observable phenomena have developed slowly. This knowledge and the earth-based data that radio astronomers have been collecting since 1954 was the basis for planning spacecraft observations that have contributed to the current understanding of Jupiter's environment. A review of the history of these observations and the theory to interpret them will allow us to appreciate the difficulty of characterizing a global magnetic field.

The History of Jovian Radio Observations

Any object that has a temperature greater than absolute zero will radiate energy to its surroundings. This radiation is emitted at all wavelengths or colors and its variation in brightness as a function of wavelength can be used to estimate the temperature of the object (see appendix 6). As early as the 1950s, however, astronomers had determined that the sun and some objects outside our solar system emitted radio signals that were clearly more powerful than could be expected from a hot radiating object. They realized that an alternate explanation was needed. Theoretical studies were made of possible mechanisms that could be responsible for the excess radio emission. These studies led to an increasing awareness by astronomers that polarized radio emission is the direct result of interaction of a magnetic field with charged particles.

In June 1954, when the Crab Nebula—a remnant of a supernova that exploded in 1054 C.E.—appeared to be behind the sun, B. F. Burke and K. L. Franklin, radio astronomers at Carnegie Institution in Washington, D.C., attempted to observe the radio emission from the Crab. They planned to measure the emission as the Crab was occulted by the sun and derive the absorption properties of the outer portion of the sun's atmosphere, the corona. Their observations were hampered by bursts of radio interference that they assumed were coming from the sun. But observations at a later date revealed that the source of the irregular activity did not move with the sun. At first it seemed to be related to the background stars, but closer inspection revealed that the noise was coming from Jupiter. In 1955 Burke and Franklin announced that Jupiter was emitting a variable radio signal at a frequency of 22.2 mHz (a wavelength of 13.6 m—see appendix 2). Subsequently the discoverers and several other observers established that the signal was polarized and detectable only within a narrow frequency range. Its strength was highly variable, and at times as strong as signals from the sun. Initially observers associated the noise with the Red Spot and White Ovals because the period of the variability in the radio noise was similar to the rotation period of these well-known atmospheric features. This idea was discarded, however, when observers noted that radio bursts were detected when none of the atmospheric features was visible. As the observations continued, it became apparent that their source rotated faster than the Red Spot, shifting eastward about a quarter of a degree per day relative to the spot. (See appendix 3 for definition of the radio period.) Later the observers noted that the strength of the signals increased when Io revolved across their field of view, further discounting the idea that the noise was generated in the clouds. But the fact that radio astronomers had absolved the jovian cloud features did not solve the problem. Instead it introduced intriguing new questions about the role of Io in this situation.

Investigations continued into the era of modern spacecraft. Sampling near-earth space allowed us to gain an understanding of the manner in which the magnetic field of a planet interacts with the solar wind. The solar wind is an outward flow of high-energy particles ejected from active regions associated with sunspots. Gases associated with the active regions are hot and ionized, and the ejected particles consist mainly of electrons and hydrogen nuclei, or protons. As these ionized high-velocity particles flow out from the sun, they are first constrained by the sun's magnetic field. But as they approach a planet, their motion is dominated by the magnetic field of the planet. Since 1958 observations with earth-orbiting satellites have revealed that the density of particles flowing out from the sun is variable and that the interaction of the solar wind with the earth's magnetic field is complex, producing a variety of phenomena. Interaction of Jupiter's field with the solar particles should generate similar phenomena.

In addition to the signals observed by Burke and Franklin at 13.6 m, D. Morris and G. L. Berge reported in 1962 that Jupiter also radiated strong signals at shorter wavelengths. Using existing receivers that had been designed for probing the interstellar region, they found that Jupiter was emitting at 3 and 70 cm. When Morris and Berge assumed that the 3 cm radiation was from a thermal source, they found that a temperature of about −130°C was indicated. This was a reasonable temperature for the cloud cover of a planet five times farther from the sun than our own planet. When they made a similar assumption for the 70 cm radiation, they found that a ridiculous temperature near 50,000°C was required. This was definitely not a thermal signature. The fact that the emission was polarized indicated that it was generated by a process involving the magnetic field.

Observations continued, and by the mid-1960s radio astronomers were aware that Jupiter's emission at infrared and radio wavelengths consisted of three distinct components: thermal emission from a cold atmosphere in the infrared and short radio wavelengths region of the spectrum; a continuous but variable signal at decimeter wavelengths (in practice this range covers wavelengths of centimeters or tens of centimeters); and a sporadic decameter signal (tens of meters) that depended on the jovian longitude and the position of Io. Such a wide range of signals indicated that the investigators were observing several different phenomena.

General Processes Occurring in Near-Planet Space

In order to understand the implications of radio emission, we must consider how charged particles become trapped in a magnetic field and why and how they emit radio signals. The fundamental theory for understanding particle entrapment was developed at the beginning of the twentieth century by Hendrik A. Lorentz, in Haarlem, Holland. His formulation, known as the Lorentz law, describes the force that a magnetic field exerts on a charged particle. This relationship can be used to investigate the path that a charged particle follows as it enters the magnetic field of a planet. Lorentz's law states that the force is *perpendicular* to the plane that contains the velocity and magnetic vectors. The magnitude of the force depends on the charge of the particle, the strength of the magnetic field, and the velocity of the particle.

The key to understanding the path that a charged moving particle must follow when it encounters a magnetic field is to realize that the Lorentz force will always be perpendicular to the direction of motion of the particle. Thus, step by step in time, the particle will be forced to move in the direction of the Lorentz force; but as it does so, the force will rotate with it. The result is that the particle must follow a curved path. In addition, if any component of the motion of the particle lies along the direction of the magnetic field, the particle will not move in a circle but will follow a spiraling or helical path along the magnetic field line.

If a planet had a magnetic field that was similar to that of the earth, the magnetic field near the equator would be nearly constant and would be aligned in a north-south direction. As a charged particle moves from this region toward either pole, it encounters a magnetic field that converges toward the poles and increases in strength. At this point in our effort to understand the nature of particle-field interaction we encounter another basic idea. The Lorentz force can change the circular component of the velocity but it cannot alter the total kinetic energy of the particle. Because the energy depends on the total velocity, if the circular component increases, the component along the magnetic field must decrease. This means that as a particle approaches one of the poles, its north-south velocity drastically decreases. The density of particles increases at the poles, causing a traffic jam. This leads to collisions of the particles in which energy is exchanged, slowing or changing the direction of some of the particles. Some particles move back toward the equator, while others collide with and transfer energy to neutral atoms or molecules in the upper atmosphere near the poles. Many of the atoms are ionized and molecules are torn apart, frequently leaving the fragments in excited states. When the atoms and molecules return to their normal state, they emit light in the form of aurorae in the polar regions.

Observations of the near-earth environment reveal that because the north-south velocity component is smallest near the pole, the charged particles trapped in the earth's magnetosphere spend most of their time in tight spirals near so called mirror points. They pass quickly through the equatorial region on their way to the other pole. In addition, they are trapped by the magnetic field and forced to rotate around the planet with it. A similar behavior would be expected when charged particles from the sun interact with Jupiter's magnetic field. The question remains as to why these particles emit radio signals. To answer this question, we must dig deeper into particle-field interactions. Here the physics is well beyond the level of this discussion, but careful consideration of the observations and theory will allow us to appreciate how fields and particles experts have interpreted the data.

Decimetric Radiation

A thorough examination of Jupiter's decimetric emission (centimeter to meter wavelengths) revealed that the wavelength dependence of the radio emission matched that of synchrotron radiation. Synchrotron radiation has been produced in high-energy experiments in the laboratory where electrons have been accelerated to velocities approaching the speed of light. Observations reveal that signals from Jupiter are highly polarized and are constrained to a narrow beam. When physicists utilize their knowledge to explain these data, they are forced to conclude from the evidence that charged particles with extremely high velocities are interacting with Jupiter's magnetic field.

The deductions are based on the fact that a high-velocity charged particle entering Jupiter's magnetic field will be accelerated so that the direction of its motion perpendicular to the magnetic field vector will be constantly changing. Laboratory experiments reveal that high-velocity electrons emit electromagnetic radiation when they are accelerated. The intensity of this radiation differs from that of a thermal source where the intensity decreases with increasing wavelength. Instead, the intensity of this polarized synchrotron radiation increases at longer wavelengths.

The manner in which the strength of the jovian signal varies as a function of wavelength is consistent with synchrotron emission from particles that are moving at very high speeds. For this case, electromagnetic theory predicts that the energy will be emitted in a narrow cone centered on the velocity vector of the particle and will be polarized in a direction perpendicular to the direction in which it is emitted. Because the charged particles trapped in a dipolar field spend most of their time in tight spirals at high latitudes, the emission will be concentrated near the poles and it will be beamed in a direction nearly parallel to the magnetic equator. Therefore the signal recorded by the earth-based observer will vary in intensity as the planet rotates, as the rotation will drive the magnetic equator in and out of alignment with our line of sight and impose a 9 hour 55 minute variation in signal strength—explaining the observed variation in the decimetric signals.

In 1960 V. Radhakrishan and J. A. Roberts showed that Jupiter's decimeter signals were emitted from a region with a radius three times larger than that of the planet. Morris and Berge provided additional evidence that the shape of the field was similar to that of the earth. They determined that the polar dimension of the emission region was less than the equatorial dimension and that the degree of polarization varied as the planet rotated. Careful consideration of the wavelength distribution of the signal led the observers to conclude that the strength of the magnetic field was at least ten times greater than that of the earth's field. Detailed analysis of the nature of the polarization when the north or south rotational pole of the planet was visible revealed that the north and south magnetic poles were reversed when compared to the earth and that the axis of the magnetic dipole was tilted about 10° relative to the axis of rotation of the planet.

These observers were convinced that they had evidence that Jupiter's magnetic field was similar but stronger than that of the earth. But they were left with a perplexing problem. To explain all the observed phenomena, they needed a supply of high-energy electrons. Because the trapped particles would lose energy in the process of emitting the radio signal, the observed amount of energy could not be generated for any extended period of time unless there was a reliable source of high-energy particles. The solar wind would supply some high-energy electrons, but not enough. The investigators hoped that spacecraft observations would reveal an additional source. At the same time planetary scientists thought the magnetic field must be generated within the planet. Could investigation of the magnetic field lead to a better understanding of the interior of the planet? In chapter 14 we will deal with this latter problem, and in chapter 15, consider how the data obtained from the Voyager Mission has influenced the investigators' thinking about field and particle interactions.

♃ Chapter 14

Jupiter's Magnetosphere

As astronomers learned more about Jupiter's magnetic field, it became apparent that the observed signals originated in the region surrounding the planet. But because the earth's magnetic field originates within the planet, they considered the extent to which observations of the phenomena associated with Jupiter's magnetic field could be used to probe its internal structure. The background for this work had already been established in the latter half of the nineteenth century, and considerable effort had been expended in an attempt to understand the magnetic fields of the earth and sun. A review of the current understanding of how magnetic fields are maintained and of the nature of the resulting magnetosphere will provide insight into the significance of the Pioneer and Voyager observations.

Generation of a Magnetic Field

In 1860 James Clerk Maxwell, an English physicist, presented a description of the interaction of magnetic and electric fields. His equations can be reduced to a single equation, called the dynamo equation, which describes the relationship between a magnetic field and the velocity of electrons moving in a region that can conduct electricity. This equation has two parts. One part describes how a field is generated in a conductive region: the magnitude of this part depends on the velocity of the moving material and its ability to conduct a current. The second part describes how the field decays with time: it depends on the ability of the material to retain a field once it has been generated. The strength and shape of the external field of a planet at any given time will be the sum of the contributions from all depths within it.

One way to understand how Jupiter's magnetic field is generated is to try to construct a self-consistent computer model of the structure of the interior of the planet. Using the model, the investigator can compute the expected magnetic field that ground-based or spacecraft instruments would measure if they could observe the model and then compare it with observations of Jupiter, working to get consistent results. We have already considered this problem in chapter 4 and have realized that many variables must be included in a model. Model builders must specify these parameters throughout the interior of the planet, and the uncertainties concerning the values of these parameters and the structure of Jupiter's interior rise to plague them. Because the dynamo equation describes the interaction between the velocity of charged particles and the magnetic field, a useful model must include a detailed description of how the velocities vary throughout the conductive portion of the planet. In addition, the ability of the material to conduct electricity must be specified over a range of temperatures and pressures, many of which are not easily attainable in the laboratory. On the positive side, however, if spacecraft can be used to better characterize Jupiter's magnetic field, the modelers can compare the observations with similar quantities predicted by their models and ultimately improve their models and their understanding of Jupiter's interior.

Results from relatively simple computer models reveal some basic characteristics of magnetic fields within rotating planets. In the case of a planet with a dipole field (like a bar magnet), the magnetic field lines extend from pole to pole and are perpendicular to the equator of the planet. If, as in the sun, the planet is rotating, then the conductive material is moving perpendicular to this imbedded magnetic field, and if the charge density is sufficiently large, the rotating material can drag the magnetic field lines along with the planetary flow, deforming the magnetic field and wrapping the force lines around and around the planet and building up a pattern that will tend to circle it parallel to the equator. This will form a ring, or toroidal pattern of the field lines, about the axis of rotation in each hemisphere and generate a component of the magnetic field that is perpendicular to the original dipole. A situation like this would generate a field that deviates from a dipole in shape. But the nature of a magnetic field is such that the farther the observer is from the source of the field, the less detectable the non-dipolar components become.

It is not clear whether the magnetic field of Jupiter would be formed deep in the planet, where the hydrogen-helium mixture has metallic properties, or nearer the surface, where molecular hydrogen could play the role of the conductor. Consideration of the models reveals that if the field were formed in a region near enough to the surface that convection could occur, small-scale convective centers would form and rise radially outward toward the planet's surface, they would distort the magnetic field, carrying the imbedded field to the surface. This predominantly radial motion would tend to reorganize the magnetic field in a manner that would restore the dipole field, masking any toroidal component that had been generated at deeper levels in the planet.

The process that generates the toroidal component is called the omega effect, while the one that restores the poloidal field is called the alpha effect. Because the time scales involved in these processes and their relative magnitudes depend on the physical state of the interior of the planet, detailed measurements of spatial and temporal variations of the external magnetic field could provide information concerning the nature of Jupiter's interior.

Numerical computer models can also be used to gain some insight into how the observations should be planned. The models show that at large distances any of the models will produce a dipolar field. Instruments will not be sensitive to departures from the dipole shape or be able to sense anomalous bumps in the field until the spacecraft passes near the planet. This constraint, coupled with the threat of damage to electronic components, must always be weighed against the other goals of a mission when a trajectory is selected.

A Planetary Magnetosphere

A rough model of Jupiter's magnetosphere, the interactive region around the planet, can be obtained by assuming that the

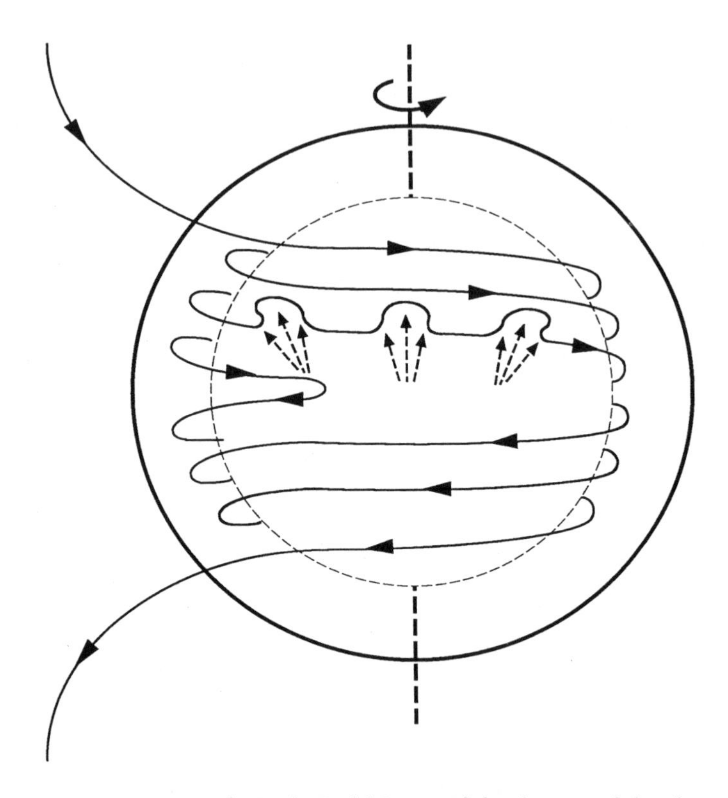

Figure 42. Distortion of Dipole Field Lines. If the density of the charged particles is large, the particles will drag the magnetic field with them as they rotate about the planet, winding the magnetic field lines into a toroidal coil. If radial convection is present, the outward motion will carry the field lines upward to the surface, where they will form loops that will contribute to a dipole component.

magnetic field is dipolar. The magnetic forces would decrease rapidly as the distance from the generating region increases (1/distance^3). Even so, the field strength decreases less rapidly with altitude than the density of the planet's atmosphere decreases with height, leaving a region above the effective limit of a planet's atmosphere that is dominated by the magnetic field. Within this region, called the inner magnetosphere, the magnetic field is so strong that it drags the charged particles with it and radiation belts form containing trapped charged particles.

Still farther out, there is a region where the high-speed particles in the solar wind will encounter the planet's magnetic field and the planet's field will act on the charged particles, much as the nose of a supersonic aircraft would interact with gas in our atmosphere. This produces a large shock wave that generates turbulence, leading to collisions among the particles, increasing the local random motion or kinetic temperature. Inside this boundary, or bow-shock, the interaction of the particles and field is dominated by the strength of the planetary field; outside the boundary, the solar wind dominates. At a given point in time the location of this interface, which is called the magnetopause, will depend on the level of solar activity.

Not only will the position of the bow-shock vary with time, but also the magnetosphere will not have a dipolar shape. Because of the impinging solar wind, the magnetosphere will be depressed inward toward the planet on the sunward side. On the opposite side the solar wind will carry the field with it, sweeping

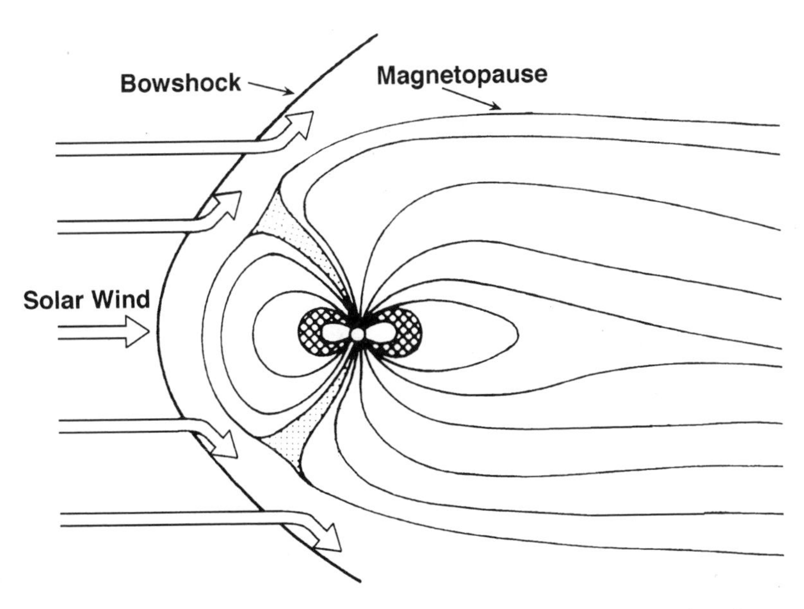

Figure 43. General Structure of a Magnetosphere. Close to the planet the field is dipolar, but farther out it is distorted by interaction with ionized particles in the solar wind. (Adapted from NASA SP-479)

the field outward to form a tail streaming out for hundreds of millions of kilometers in the anti-solar direction.

When a spacecraft enters the vicinity of a planet that possesses a sizable magnetic field, the data obtained with the directional particle detectors will change as the spacecraft approaches the planet. First, the direction and distribution of the particles are determined by the direction and density of the solar wind. Then the distribution will change to chaotic motion in the region of the bow-shock. After the ship has crossed the bow-shock, the data will be strongly modified by the magnetic field of the planet.

Near the bow-shock, at the interface between the interplanetary region and the magnetosphere, a major part of the streaming motion of the electrons and protons in the solar wind is converted to turbulent motion. When the average speed of particles involved in this chaotic region is interpreted as a kinetic temperature (proportional to mass $\times$ velocity^2), very high temperatures are reported. But the degree of spacecraft heating depends on the *rate* of energy transferred to the spacecraft. This rate will depend on the number of impacts per second that the craft experiences. Because the local particle density is very, very low, the number of impacts will be so small that the spacecraft can easily radiate away the energy that it absorbs. But even though there is no danger that the spacecraft would melt, there will be damage. High-energy electrons can readily pass through the outer layers of the spacecraft and cause direct damage to the electronic components of the instruments and computers. Therefore, while the spacecraft is in the vicinity of the planet, it accumulates damage that can cause serious malfunction. The need to avoid this damage imposed constraints on the Pioneer and Voyager mission plans.

Considerations for Mission Planning

The knowledge that investigators had gained influenced both the trajectories of the Pioneer and Voyager missions and the types of instruments that were selected for their payloads. By the early 1970s the temporal variation in the radio data had been used to establish the rate of rotation of the core of the planet. Interpretation of more than a decade of decimetric data led to predictions that the magnetic field was similar in shape to that of the earth but that the strength of the field at the cloud tops was ten times greater than the magnetic field at the earth's surface. Because Jupiter has a large mass, an internal heat source, and a rapid rate of rotation, it was generally accepted that a portion of the interior must be conductive and the internal heat source should drive a large enough internal circulation to support a healthy magnetic dynamo. It was not certain, however, at what depth within the planet this would occur. Thus investigators hoped that they could obtain more knowledge about the structure of the field to help them determine how well the current models described the internal structure of the planet. To do this, they wanted to sample the particles and field at many different locations and as many times as possible.

Systematic variations in the strength of the decametric data indicated that the revolution of Io, the nearest Galilean satellite, imposed its own modulation on the signal. Various models of the interaction of the satellite with the magnetic field encouraged close inspection of both the inner magnetic field and the environment near Io. Armed with predictions of the type of magnetospheres that the spacecraft would encounter, the investigators and engineers set about to design instruments for the Pioneer and Voyager spacecrafts that would sample the magnetosphere over a range of distance and in different directions relative to the impinging solar wind. Although the observable signals would be generated above the cloud deck, the pattern of observed emission could be used as a probe to learn more about Jupiter's interior.

♃ Chapter 15

Pioneer and Voyager Observations of the Magnetosphere

In the past twenty-five years scientists have become increasingly aware of the danger of exposing astronauts to high-energy particles during extravehicular activities. Such awareness has caused them to study how the earth's magnetic field shields us from damaging particles that are ejected from the sun and has increased their interest in sampling the near-planet environments of other planets. Although radio astronomers were aware of the global scale of the earth's magnetic field, it was only when spacecraft flew by the other planets that they could fully appreciate the wide range of magnetic fields that exists in the solar system.

Because the ionized particles in the solar wind disperse more or less uniformly as they move outward from the sun and because Jupiter is five times farther from the sun than the earth, the particle density of the solar wind is considerably less at Jupiter than at the earth. This fact, and consideration of Jupiter's larger magnetic field, led spacecraft design teams to expect that Jupiter would have a more extended magnetosphere. The investigators hoped to gain new insights into the mechanisms that produce the radio signals described in chapter 13.

The Pioneer 10 and 11 spacecraft, which encountered Jupiter in 1973 and 1974, carried seven experiments designed to probe the jovian magnetosphere. Edward Smith, from the Jet Propulsion Laboratory, Mario Acuna, from Goddard Space Flight Center, John Wolfe, from Ames Research Center, John Simpson, from University of Chicago, Frank McDonald, from Goddard Space Flight Center, James Van Allen, from the University of Iowa, and R. Walker Fillius, from the University of California at San Diego, were the principal investigators of the seven experiments. Co-investigators, staff members at the Ames Research Center, and several subcontractors participated in the design and fabrication of this battery of instruments.

The Pioneer Encounter

The orbits of Pioneer 10 and 11 were designed to satisfy a range of objectives. The trajectory of Pioneer 10 was chosen to probe the radiation belts, image the atmosphere and satellites with a spin-scan camera, and to pass behind Io so that Pioneer 10 would be occulted by it. Occultation measurements of a well-known radio signal would be a sensitive method of detecting a very thin atmosphere, therefore this experiment could provide information concerning the mechanism that generates the Io-nian modulation of the decametric radio data. The Pioneer 11 trajectory was determined largely by the requirement that the spacecraft continue on to Saturn. To gain the acceleration needed to alter Pioneer's trajectory, the spacecraft had to pass quite close to Jupiter. Whether or not the electronic components could survive the damage that this environment would inflict on them was not known.

As Pioneer 10 approached the planet, the particle detectors recorded changes in the rate of impact of charged particles that indicated that the ship had encountered the bow-shock and passed through the turbulent region into the magnetosphere. This occurred at a distance of 6.8 million km (95 planetary radii from Jupiter), or at about a hundred times greater distance from the planet than the location of a similar boundary of the earth's magnetosphere. Soon after the initial crossing, an increase in density of high-energy particles arriving in the incident solar wind caused the magnetopause to recede toward the planet faster than the spacecraft was approaching. The spacecraft was forced to cross the bow-shock and magnetopause once again on its way to its closest approach to the planet. The expanding and contracting boundary of the magnetosphere changed the shape and location of the outward streaming magnetotail, causing it to cross the post-encounter trajectory of the spacecraft seventeen times. This illustrates the sensitivity of the location of the boundary to the level of solar activity. Pioneer 11 encountered the magnetopause three times on the inbound trajectory as well as three times on the outbound path.

Analysis of the Pioneer data confirmed that the equatorial strength of the magnetic field at the level of the cloud tops was about 10 gauss (at 45° north latitude the north-south component of the earth's magnetic field is about 0.2 gauss) and could be described in general as a dipole, with the magnetic axis tilted 11° relative to the axis of rotation of the planet. The data were in excellent agreement with the results that had been obtained from the ground-based observations of synchrotron emission. The fact that the polarity of the jovian magnetic field is reversed relative to that of the earth was also confirmed (our north-seeking compasses would point south in the vicinity of Jupiter). Although the present orientation of the magnetic field is of interest, it is not disturbing: geological evidence shows that the direction and strength of the earth's magnetic field is not constant, and sunspot observations reveal that the sun's magnetic field undergoes a cyclic variability, reversing polarity from north to south and back again to north in a twenty-two year period.

When measurements of the strength and direction of the magnetic field were compared with those predicted by an ideal dipole field, it was apparent that the magnetic field near the planet was more complicated than that of the earth. Because this sort of detail becomes harder to detect at larger distances, the investigators interpreted it to indicate that the region that generates Jupiter's magnetic field is not the inner core but must be located in the region between the dense core and the atmosphere. A phenomenon described as the omega effect (see chapter 14) would be generated in the lower portion of the convective envelope by the interaction of rising cells and motion about the axis of rotation. At these depths hydrogen is dense enough to display metallic properties and the conductivity would be high enough to generate additional structure in the magnetic field.

The Pioneer data also revealed that the population of particles in the radiation belts was high and that bursts of electrons were released into space. The bursts tended to occur at ten-hour intervals and seemed to originate from complex interac-

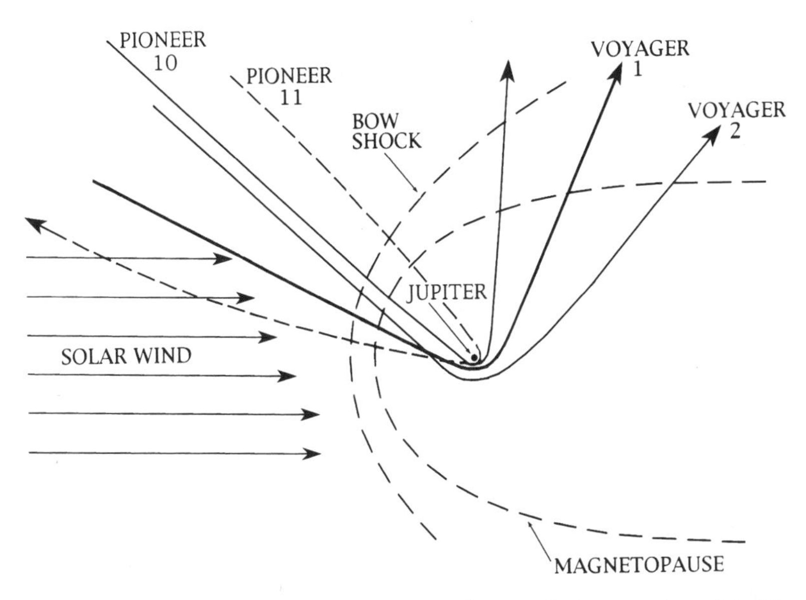

Figure 44. Trajectories of Pioneer 10 and 11 and Voyager 1 and 2. The Pioneer spacecraft passed through the inner magnetosphere and the Voyagers passed farther from Jupiter, sampling the outer magnetopause. (Adapted from NASA SP-446, Ames Research Center)

tions within the radiation belts. Electrons with energies in the million electron volt range (1 electron volt, or 1 eV, is the energy that an electron gains if it is accelerated across a voltage difference of 1 volt) were detected in the outer radiation belts. Energies that would be expected from solar wind particles trapped by the magnetic field (in the thousand electron volt range) fall far short of the observed values. These observations confirmed the suspicions of the earth-based astronomers, but the investigators had no explanation of how these huge velocities could be reached.

The Pioneer spacecraft revealed another interesting aspect of Jupiter's magnetic field: the satellites from Ganymede inward, including Europa, Io, and Amalthea, lie inside the magnetosphere. As the satellites orbit Jupiter, they can sweep up or interact with the charged particles trapped in the radiation belts. The action of the large satellites can create regions that are stripped of the ionized particles. If their outer layers are conductive, on the other hand, they can change the pattern of electrons moving along the magnetic fields and alter the current as they pass by. Pioneer 11 instruments measured such a phenomenon in the field linking Io with Jupiter. The radio scientists proposed this as the source of the sporadic decametric signal that had been observed with earth-based telescopes. These results influenced planning for the Voyager 1 and 2 observing sequences, and efforts to study this phenomenon, called the "Io flux tube," were given high priority.

The Voyager 1 and 2 Encounters

Each of the Voyager spacecraft carried six instruments for measuring magnetic field strengths, particle densities, and electromagnetic wave phenomena. The instruments were designed to make direct measurements of the magnetic field and the spatial and temporal dependence of the density and velocity of various ions, and to detect and analyze signal bursts and electromagnetic wave phenomena. The experiment teams included several scientists from Germany and France who, along with the other co-investigators, had proposed as individual teams with a principal investigator for each experiment.

Norman F. Ness, of NASA Goddard Space Flight Center, continued the investigation of Jupiter's magnetic field that he and Mario Acuna had begun with the Pioneer instruments. With a team of five co-investigators, he operated a pair of magnetometers. One was designed to measure strong fields and the other was sensitive to weak variations in the field. This second magnetometer was also sensitive to variations in the operations of the spacecraft. In order to overcome this source of interference, the instrument was mounted on a boom that was folded in a canister during launch. When it was deployed, it extended outward 13 meters from the spacecraft. By adjusting the sensitivity of these two magnetometers and taking advantage of their spatial separation to eliminate transient fields associated with spacecraft operation, magnetic field strengths spanning six orders of magnitude (with the strongest fields 100 times that at the earth's surface) were measured.

Herbert S. Bridges, from the Massachusetts Institute of Technology, and eleven co-investigators designed an instrument to measure the densities and temperatures in the inner and outer regions of the magnetosphere and in the tail that extends outward in the anti-solar direction. A complementary instrument was constructed by Stamatios M. Krimigis, a physicist from Johns Hopkins University, and his team of six co-investigators. Their instrument had two detectors designed to measure high-energy electrons (in the 10 thousand to 11 million eV range) and protons (with energies ranging from 15 thousand to 150 million eV) as well as nuclei of helium and even heavier atoms. One detector was constructed to work in a low-density environment and the other, to sample the higher density region close to Jupiter. The instruments were mounted on their own stepping platform and could be pointed in eight different directions, with a stepping interval that could be varied from 48 seconds to 48 minutes to allow variable sampling of the environment. The goal of this experiment was to determine the energy and mass of incident particles and to allow the team to determine whether the particles were electrons or the nuclei of hydrogen, helium, or some other atomic species.

The planetary radio astronomy instrument, which was designed to study the nature of the decameter and kilometer signals, consisted of two antennae that extended outward from the ship for 10 meters. James W. Warwick, from the University of Colorado, and his eleven co-investigators operated the receivers in several modes, one of which could detect rapid variations in the radio emission.

When a gas is highly ionized, it is called a plasma. The plasma will probably have a balance of positive and negative charges, but because the masses of the positively charged hydrogen and helium nuclei are 1800 to more that 7000 times more massive than the negatively charged electrons, waves can be generated that can be observed as variations of an electrical field. The team that expected to detect these plasma waves was headed by the late Fredrick L. Scarf, of the TRW Defense and Space Systems Group of Redondo Beach, California. Their instrument used the 10-meter antenna and measured variations in the electric fields. Scarf delighted attendees of the encounter press conferences when he converted his data to an audio tape to illustrate the phenomena that he was measuring.

An additional instrument was included on the spacecraft that was designed to measure particles of extra-solar system origin. Although Rochus E. Vogt, from the California Institute of Technology, and his six co-investigators operated the instrument in the vicinity of Jupiter, the chief interest of the team was to continue measurements outward through the solar system and to map the interaction between the solar wind and high-energy or cosmic-ray particles from deep space. This study is continuing, requiring only an occasional read-out through the Deep Space Network.

Coordinated efforts between the fields and particles experiment teams, as well as careful negotiations with other groups with conflicting goals, were required to optimize the science yield from the two Voyager encounters. Much of the planning was based on ground-based investigations and the Pioneer results. Although this earlier information had given investigators a wealth of questions in need of answers, they had to face the fact that the trajectories selected for the two flybys were those that optimized the goals of the whole mission. A desire to achieve a close pass by the large Saturnian satellite Titan and to preserve the option for Voyager 2 to continue on to Uranus and Neptune placed strong constraints on the trajectories of the two spacecraft. Where Pioneer 10 and 11 passed within 2.8 and 1.7 planetary radii, respectively, of Jupiter, Voyager 1 would approach to within about 5 planetary radii and Voyager 2 would come no closer than 10 radii. As a result, the Pioneer encounters yielded information on the near-planet environment; the Voyagers added to the knowledge of complex processes occurring in the outer magnetosphere but provided less information about the internal structure of Jupiter.

As Voyager 1 approached the planet it crossed the bow-shock at a distance of 85.7 planetary radii. As with the two Pioneer encounters, the variable solar wind, buffeting the sunward extension of Jupiter's magnetosphere, generated five more bow-shock crossings on the in-bound leg of the Voyager's journey. Near the magnetopause energy from impacting particles absorbed by the Voyager instruments indicated that particles were moving at speeds equivalent to what they would have at temperatures of 300–400 million degrees C. Although very high-velocity particles were impacting the spacecraft, their density was too low to cause noticeable heating of the surface. However, these particles did present a radiation hazard for the on-board electronics.

The Voyager results indicate that the structure of the magnetosphere is quite complex. For example, out beyond the orbit of Ganymede but still inside the bow-shock, a concentration of charged particles orbit the planet in the plane that contains the magnetic equator. Because the magnetic equator is tilted about 11° relative to the rotational axis and the charged particles are constrained to move with the field, this disk of charged particles swept up and down relative to the spacecraft and transmitted a variable signal to the instruments as the planet rotated. Within the frame of reference of the spacecraft the passage of this current sheet was reminiscent of a great badly mounted floppy disk.

On March 5, 1979, the Voyager 1 spacecraft passed 20,000 km south of Io into the region where radio scientists expected to encounter variations in the concentration of charged particles that they thought were associated with the decametric bursts. Although no change in the intensity of the magnetic field was measured, a significant change in the direction of the field was observed. The investigators interpreted the characteristics of the signal to indicate that Io has no magnetic field; instead, they proposed that the spacecraft had passed near a current system. They reasoned that because Io revolves about Jupiter in 1.77 days, the trapped charged particles that are forced to co-rotate with Jupiter's magnetic field (rotating once in about 10 hours) must catch up with Io and sweep past twice a day; the relative motion of the charged particles would set up an electric field in the satellite. This would allow a current to flow through the closed circuit formed by Io, Io's extended ionosphere, currents in Jupiter's magnetic field, and currents in Jupiter's ionosphere. The resulting conducting circuit would create a flux tube. This well-known phenomenon is the basis of airport metal detectors.

The team members concluded that an object as large as Io would generate a huge amount of power and could explain their observations; also that a large amount of energy should be dissipated as the current passes through the jovian cloud deck and

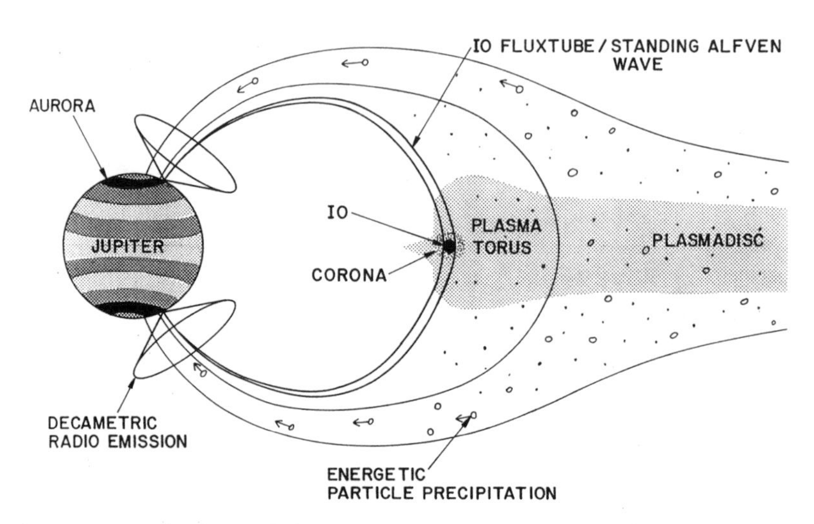

Figure 45. Io's Flux Tube. The flux tube follows field lines that emerge from the planet, pass through Io, and reenter Jupiter. As the rotating magnetic field sweeps by Io, a current is generated. (Courtesy of F. Bagenal, University of Colorado)

intersects the surface of Io. These regions may display distinctive types of activity, and therefore are desirable competitors for the rationed *Galileo* resources.

Clouds of Atoms Associated with Io

In 1973 Robert Brown, then at Harvard, reported that he had detected yellow emission from neutral sodium atoms near Io. Before the arrival of the Voyager spacecraft other ground-based observers established that singly ionized sulfur and oxygen were also present. As a result of this earlier work the wide-angle camera was equipped with a narrow yellow filter to image the distribution of neutral sodium atoms. Although these observations confirmed the ground-based results, the Voyager UV spectrograph yielded more significant data. In keeping with the earlier ground-based discoveries, the UV investigators determined that the cloud was mainly of oxygen and sulfur. This information, combined with the observations of active volcanism, was consistent with a sulfur-driven volcanism, raising much more interest in the distribution of various ions of sulfur, observable in the UV, than had been anticipated.

Both the sporadic nature of volcanic activity on Io and the variability of the solar wind will cause this cloud to vary greatly with time. With this in mind and because the yellow light from neutral sodium can be observed with ground-based telescopes over a large area around Jupiter, Nicholas Schneider, at the University of Colorado, and collaborating scientists at the University of Arizona have developed movies that reveal how the density of the cloud varies with time. Schneider's success assures us that even though the current data rate from the Galileo spacecraft is inadequate to pursue this problem, an ongoing cost-effective study, using modern imaging techniques, can be pursued with ground-based telescopes.

Although atoms ejected by the volcanoes on Io can leave the surface and escape from the satellite, few have sufficient energy to escape from Jupiter's gravitational grasp. The result is that they form a cloud of particles orbiting Jupiter in the vicinity of Io. Any particles in the area that are ionized will be pulled along by Jupiter's rotating magnetic field, while neutral ones will not; this will enhance the possibility of collisions between the two types of particles. These collisions, along with exposure to solar ultraviolet radiation, will tend to increase the level of ionization and determine the degree to which the particles will interact with the magnetic field. The charged particles will disperse from Io and form a doughnut-shaped distribution, or torus, of ions that will be centered about Jupiter's magnetic equator. This torus is inclined relative to Io's orbital plane, which is in the plane of the rotational equator of the planet.

The ultraviolet spectrometer and the cosmic-ray instruments on board the Voyagers recorded the presence of enhanced populations of oxygen and sulfur closer to Jupiter, indicating that particles that escape from the torus spiral inward. Thus they may be the source of some of the high-energy particles in the inner magnetosphere. Analysis of Voyager 2 data yielded ultraviolet emission stronger by a factor of two than it had been four months earlier during the Voyager 1 encounter. In addition, the

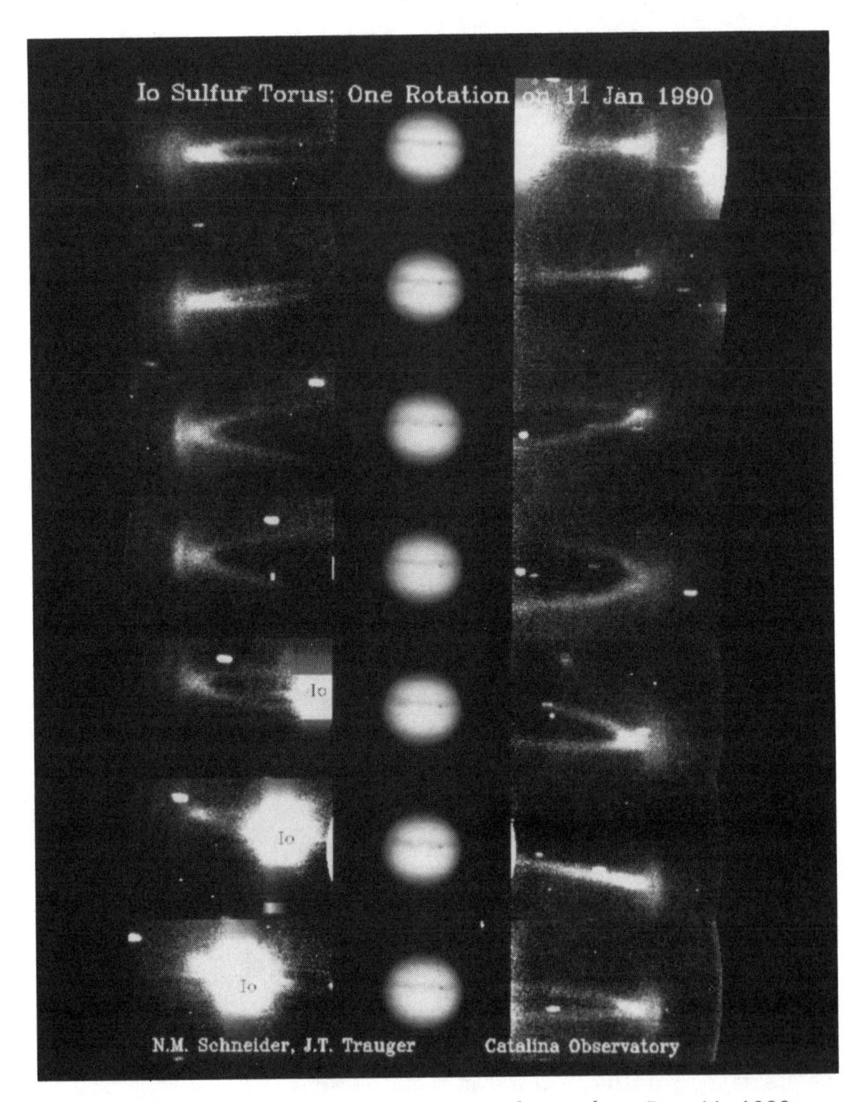

Figure 46. Io Torus. A series of images, obtained on Jan. 11, 1990, at Catalina Observatory, showing the motion of the Io Torus as Jupiter rotates. The images of Jupiter have been superimposed on a mask used to prevent overexposure of the original images. (Courtesy of N. Schneider, University of Colorado, and J. T. Trauger, JPL, NASA)

kinetic temperature had decreased by 30 percent, indicating that the particles were moving more slowly.

An explanation of variations in these quantities over such a short time span could be that there is not a large amount of mass in the torus and that the replenishment rate from the satellite, with its variable eruptions, must nearly balance particle losses due to interaction of ions with the field. If this is the case, the torus is variable and complex. It is regrettable that the limitations on the *Galileo* data rate will restrict observations of this region.

Future Investigations

The field of planetary magnetism has developed rapidly with the advent of space exploration. Although the Pioneer Mission confirmed the results that had been obtained from ground-based observations concerning the inner magnetosphere and revealed many details about the middle and outer regions of the magnetosphere, the mission raised many questions. The Voyager Mission added additional data concerning the temporal and spatial variability of complex structures associated with the magnetic field. In addition, this mission has provided a valuable sampling of the structure of the magnetic field of the other outer planets. The *Galileo* observation will yield additional data. But constraints that require the spacecraft to orbit beyond the orbit of Io in order to avoid radiation damage to the components of the ship will severely limit information about the inner magnetosphere and core of the planet. Even if the transmission problems were solved, the need to maintain an equatorial orbit to assure close encounters with the satellites would further limit the magnetic field data.

The results from Voyager and the limited expectations for *Galileo* make it apparent that a multipurpose spacecraft is not the best choice for studying Jupiter's magnetic field. Planetary orbiters that map the three-dimensional space for many radii around the planet are needed to acquire data sets that provide a longer period of consistent sampling and allow scientists to develop more realistic models of the mechanisms that generate and maintain the magnetic field. This could be accomplished by inserting a spacecraft into a very eccentric orbit that is highly inclined to the equator of the planet. The spacecraft's orbit could then be altered so that it moved progressively around the planet relative to the incoming solar wind. If this were done, the instruments could sample far and near space at many latitudes and longitudes. A spacecraft in an eccentric, inclined orbit could dive through the radiation belts at high speeds, nearly graze the atmosphere, and race back out to a safe distance. Not only would such a mission do an excellent job of sampling the magnetic field, but it could also provide a more complete sampling of the gravitational field. This sort of scheme was examined by a working group of planetary scientists headed by George Siscoe. The participants in this workshop, held at University of California at Los Angeles on July 18–19, 1985, concluded that this approach was feasible with current technology. Implementation is not included in plans for missions in the near future, however. Instead, ground-based observations are continuing as we anticipate *Galileo's* 1995 arrival at Jupiter.

♃ Part V

CONCLUSION

♃ Chapter 16

Future Observations of the Jovian System

In March 1993, as team members prepared for *Galileo's* flyby of the asteroid Ida and its encounter with Jupiter, other astronomers made an interesting discovery. Using the 46-cm (18-in) Schmidt camera at Palomar Observatory, Eugene and Carolyn Shoemaker and David Levy found a peculiar elongated object. In the days that followed, careful analysis of its path among the stars revealed that this strange object was a comet that had become trapped by Jupiter and in July 1992 had passed so close to the planet that it had been torn into twenty some odd pieces.

As time went by and the orbits of the pieces became better defined, Donald Yeomans, Paul Chodas, and Zdenek Sekanina, at the Jet Propulsion Laboratory, found that the pieces would smash into the planet. They predicted that the collisions would occur on the far side of the planet during a period of several days centered on July 19, 1994. This once-in-recorded-history event fired the imagination and whetted the curiosity of planetary scientists, and they proceeded to set up an international network to record the effects of these collisional events.

Complementary to this exciting event, the December 1993 refurbishment mission of the Hubble Space Telescope and the arrival of *Galileo* at Jupiter two years later will provide new tools to answer a few of our questions. In this chapter we will review the long-term investigations that are planned with *Galileo* and the Hubble telescope and acknowledge new technology that allows us to observe the planet with large ground-based telescopes in infrared light. Chapter 17 discusses the comet.

Although the unprecedented success of the Voyager flybys completely altered the way the public views the outer solar system, many questions were not answered. It had been recognized that this would be the case even during the earliest planning stages of the Voyager Mission. At that time the scientific community was divided as to whether the Voyager survey or an intensive investigation of the jovian system should be given priority. Therefore both types of missions were studied in parallel; NASA had decided to construct the Galileo spacecraft before the Voyager arrival at Jupiter. In addition, the Cassini spacecraft, scheduled to arrive at Saturn after the turn of the century, will fly by Jupiter and could collect another data set. Thus four limited spacecraft-acquired data sets—Pioneer (1973–74), Voyager (1979), *Galileo* (1995–97), and *Cassini* (post 2000)—could become available.

The Galileo Trajectory and Associated Problems

Even though preliminary study of a Galileo-type mission began in the early 1970s, budgetary constraints and the Challenger accident, with subsequent rescheduling of shuttle launches, conspired to delay launching of the Galileo Mission until October 18, 1989. Having missed a launch window that would have allowed a direct interplanetary trip, the engineers devised a plan to take advantage of multiple gravitational assists from inner planets. By launching the spacecraft so that it passed just behind Venus early in 1990, the spacecraft was accelerated and placed on a trajectory that brought it back by the earth in less than a year. It then swept out into the asteroid belt, orbiting the sun and coming back by the earth a second time. By using carefully planned maneuvers during the second earth flyby in December 1992, the engineers adjusted the angle of approach and the miss distance from the earth to gain speed and to control the direction of the spacecraft so that it will finally arrive at Jupiter in December 1995.

Before this plan—dubbed the VEEGA (Venus-Earth-Earth Gravitational Assist)—could be used to allow a spacecraft as massive as *Galileo* to reach Jupiter, another major problem had to be solved. The spacecraft had not been designed to withstand the more intense heating in the inner solar system. A sunscreen was developed to compensate for this, and the large high-gain

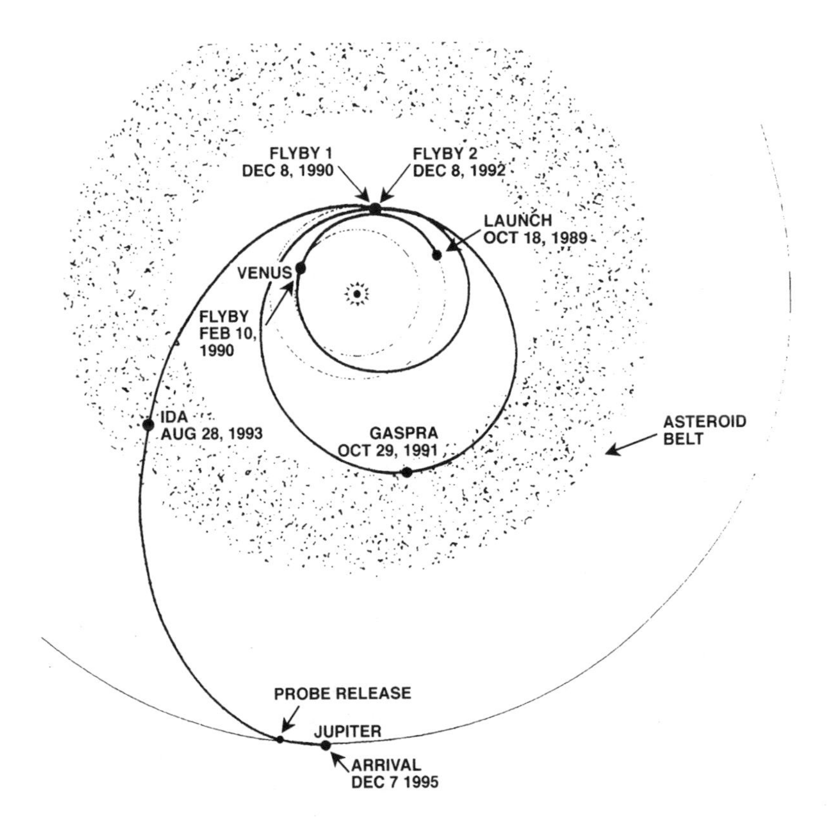

Figure 47. The VEEGA Trajectory. *Galileo's* schedule of events from 1989 to 1995. (JPL, NASA)

antenna was to remain furled until after the spacecraft had swept by Venus and was back out in the vicinity of the earth. The plan seemed adequate, but when the commands were sent to open the antenna, it did not fully extend. Small pins that secured the furled antenna remained in place on one side. This renders the antenna useless. Without it, data must be transmitted using the less efficient low-gain communications antenna and the rate with which data can be transmitted from the spacecraft is greatly restricted. Although the data from Venus, Moon, and Earth, and the asteroids Gaspra and Ida have revealed that the instruments are fully operational, there is no way the original mission design can be carried out without a fully extended high-gain antenna.

The hope of solving the problem of the jammed antenna delayed detailed planning for data acquisition during the mission. Various attempts to cool the antenna or hammer (vibrate) the extension device to dislodge the pins have failed. Even so, the mission is not lost. At low-gain antenna data rates, valuable data will be acquired. High-resolution mapping of the satellites' surfaces and the atmospheric data derived from the small entry probe will be unique. The fact that *Galileo* will arrive sixteen years after Voyager will not only provide some useful time-dependent information about the atmosphere and Io's volcanic activity but will also allow fields and particles studies at a different phase of the solar cycle, helping investigators to gain more insight into the Sun-Jupiter interaction.

Galileo will arrive on a trajectory that intersects the orbit of Jupiter at a time when the spacecraft will pass close to Io. Carefully timed firing of the retrorockets will allow the gravitational attraction of Io to pull the ship into a very elliptical orbit around Jupiter. A series of maneuvers can then be executed to trim subsequent orbits to the desired shapes and orientations, allowing the spacecraft to pass ahead of or behind selected satellites. The design of the camera allows an image to be read out in 8.3 seconds. This is faster than could have been transmitted in real time had the antenna been fully functional. This design, with on-board data-storage capability, was chosen to allow nearly grazing encounters with satellites. Such flybys are of very short duration, hence good planning and adequate data storage were absolutely necessary to get the maximum amount of information.

Even though investigators will map a much smaller fraction of the satellite surfaces, many of the goals of the satellite investigations can still be attained because of *Galileo's* read-out and storage capability. The main problem will be how to send the data home once it is acquired. The Deep Space Net, with large receiving dishes in California, Spain, and Australia, has the capability of linking with other radio telescopes to optimize the data rate at preselected times. The net has been improved since the days of Voyager and will be ready to receive the data. Even with this enhanced receiving capacity, however, the data rate is still painfully slow. The problem is simply a plugged data pipe (the unextended antenna).

Another way to address the low-gain data rate is to compress the data. A technique called cosine transform compression (using our old and powerful friend, the Fourier transform) was developed for the Mars orbiter. The Galileo team is planning to install software on the spacecraft that will perform this type of compression. This and the fact that the mission was planned so that each orbit will be about two months long, with the spacecraft far from either Jupiter or one of the satellites most of the time, will allow the investigators to obtain significant data, even at the low data rate.

Plans for the Probe

The Galileo spacecraft is carrying an atmospheric probe that will be separated from the orbiter about 165 days before the orbiter is inserted into orbit. The probe will travel along a curved path that will carry it into the atmosphere at about 6° north latitude. Six instruments will be encapsulated in a saucer-shaped pod 87 cm high and 125 cm in diameter. Because the pod will plunge into Jupiter's atmosphere at about 60 km/s, it will generate a great deal of frictional heating. The designers have dealt with this by constructing the outer layers of the nose cone from carbon phenolic, which will heat up and ablate off the pod, taking much of the heat with them. The frictional interaction of the craft with the atmosphere will reduce the craft's speed to 0.8 km/s. Deeper into the atmosphere a parachute will be deployed to further slow the probe. The instruments will then begin an intensive sampling program which will last only a few minutes before the probe descends so far into the atmosphere that it can no longer communicate with the antenna of the orbiter, which will be passing overhead before it attains orbit about Jupiter.

The probe is managed by Ames Research Center, in Palo Alto, California. The six instruments will measure temperature, density, pressure, composition, and molecular weight of the gas, sizes and scattering properties of the cloud particles, and the outward heat flow from Jupiter's interior. This is the only spacecraft that NASA is now developing that will directly sample the atmosphere of any of the outer planets (*Cassini* will probe Titan,

Saturn's largest moon). The absence of possible follow-up makes this effort especially significant to our understanding of conditions below the cloud deck of the giant planets.

Plans for the Orbiter

The Galileo spacecraft uses microcomputers to drive the individual instruments instead of depending on a central computer as Voyager did. Each instrument interfaces with the on-board data storage and the main computer is used to control which of the stored data is transmitted to the earth at any time. The spacecraft is composed of two parts that combine the desirable characteristics of the Pioneer and Voyager designs. One part, with three-axis stabilization similar to that of Voyager, houses instruments that require pointing stability to avoid smearing of images. A second part of the spacecraft, designed to rotate three times per minute, contains the instruments that measure the angular dependence of the magnetic field and impacting particles.

The orbiter carries nine instruments. Four deal with spatially resolved remote sensing or imaging and five measure magnetic fields and particle densities. The television camera uses a solid-state detector and the optical arrangement is similar to the narrow-angle camera on board the Voyager; but the detector, a charged-couple device (CCD) 100 times more sensitive, is capable of detecting red and near-infrared light.

The near-infrared mapping spectrometer can obtain spectra at wavelengths up to 5 microns, 10 times longer than the human eye can see. The instrument disperses the light perpendicular to a narrow entry slit so that a single read-out of the two-dimensional array reveals how the observed area reflects the light in as many as 200 visible and infrared colors. The spatial resolution of the television camera is 25 times better than this instrument, but the camera sees only 7 colors in visible and near-infrared light. Therefore these instruments can be used together to map the morphology of a region and to identify absorption due to sulfur dioxide, water ices, and silicate minerals on the satellites and methane and ammonia absorption in Jupiter's atmosphere.

The photopolarimeter-radiometer is sensitive to even longer infrared wavelengths and is designed to measure surface temperatures and heat flow and to determine the heat retention of different surface areas on the satellites. In contrast, the ultraviolet spectrometer probes the outer reaches of the atmosphere. It can also be used to map regions of sulfur-dioxide deposits on Io and to detect variations in the number of charged ions and the density in Io's torus and in the tenuous gases around the satellites.

The fields and particles instruments will monitor temporal variations in Jupiter's magnetosphere and measure interactions between the satellites and the magnetic fields and determine charged particle densities associated with these phenomena. The dust detector will be able to determine the rate of infall of interplanetary meteoroids into the jovian system. The extent to which this can be done will depend on the amount of data that can be steadily dribbled back over the duration of the mission.

The telemetered signal and the response of the ship can be used to gain additional information. As seen from the ground-based receiving station, the spacecraft will undergo multiple occultations by the planet or a satellite. When the occultations are considered as a group, even though the sampling rate will be slower than planned, the modified signal will yield information concerning temporal and spatial variations in the jovian atmosphere. The length of the multiple occultations can also be used to refine the measurements of the satellite diameters.

By comparing the frequency of the signals received by the Deep Space Net to the frequency transmitted by the spacecraft, the velocity of the spacecraft relative to the receiving antenna can be deduced. Careful analysis of these velocities throughout the whole mission will allow the engineers to determine the forces exerted on the spacecraft by the planet and satellites. The precise manner with which the spacecraft is accelerated depends on whether the mass of the interacting bodies is spherically distributed and whether the density is concentrated toward the center or is more homogeneous throughout the body.

The following sections review what information the Galileo spacecraft can obtain given a limited data rate and heavy use of data-compression techniques.

Observations of the Four Inner Satellites

Even without the antenna problem the exploration of the four inner satellites—Amalthea, Andrastea, Metis, and Thebe—would be fundamentally limited by engineering constraints. The spacecraft must orbit outside the region where high-energy particles are trapped in the magnetic field to avoid damage to electronic components. After the initial orbital insertion, the spacecraft will not come within six jovian radii of these satellites. At this distance the spatial resolution will be poorer than 8 km and similar to that which the Voyager cameras obtained for Amalthea. At this resolution some information about the shapes, colors, and whether the satellites rotate synchronously could be obtained.

Precise determination of where one of these satellites was located within the camera's field of view could allow a small section of an image to be sent to earth, minimizing the number of bits needed to retrieve satellite images. Because *Galileo* is an orbiting spacecraft and the camera is sensitive in the near infrared, it is possible to view the satellites from a variety of solar illumination angles and the nature of the reddened surfaces of these satellites could be investigated. These observations might determine if a satellite's surface is covered with the dark-red material or if small craters reveal a different underlying material. Because these observations must compete with others, it is doubtful whether many will be obtained. The investigators must decide which observations will receive highest priority.

Return to the Galilean Satellites

Although the insertion of the spacecraft into a large elliptical orbit requires a close pass by Io, the potential for damage due to high-energy particles is especially severe near this satellite. Therefore, during the first orbit, the engineers will command the spacecraft to reduce the major axis and adjust the orbit to a more circular shape, ruling out the possibility of more high-resolution observations of Io. Still, with limited imaging and additional heat-flow measurements with the infrared instruments, geologists may be able to determine the chemical composition of some of the structures (i.e. whether specific flows are sulfur or silicate rich) and to document patterns of activity. When the *Galileo* data are compared with the Voyager results, the time scales related to plume activity and surface repaving will be further clarified.

In comparison to Io, Europa is far enough from Jupiter that close encounters are possible and features as small as tens of meters can be resolved. Even though the geologists will be required to limit the total surface area that they map, with higher resolution multiple encounters and new information about color differences of specific types of features, it may be possible to determine the thickness of the icy crust and to understand the processes that have formed the intricate system of fine fractures. The possibility of seeing small impact craters should place added constraints on the resurfacing rates. The spatially resolved infrared observations will add to the knowledge of the chemical composition of the brown and gray regions and allow us to estimate local rates of heat flow and possibly to detect steam venting.

The *Galileo* investigators had originally planned to image Europa so extensively that they could make detailed measurements of the satellite's shape. They planned to determine its response to tidal forces imposed by Jupiter and to use these distortions to place constraints on its internal structure. But with the current data rate these goals will not be attainable. It remains to be seen whether investigators will determine if the icy crust is underlaid by a liquid water ocean or by an active silicate mantle.

Similarly, close encounters with Ganymede and Callisto will be obtained. The ability to adjust the spacecraft's trajectory to allow it to pass in front of or behind these satellites will allow selected high-resolution mapping of the leading and following hemispheres.

On Ganymede, the investigators will be able to image limited areas in the grooved regions, map the spatial arrangement of the complex surface patterns, and determine relative ages of the various episodes of activity. They should be able to acquire enough high-resolution data to consider whether the observed patterns can be explained by expansional forces alone or whether compressional forces due to motion of crustal plates are also required.

Detailed study of craters of many sizes will indicate how they have changed with time. The infrared instruments can produce maps of the composition based on infrared reflectivities and measure heat flow. These data can be correlated with the local morphology to determine the level of ice enrichment and strength and viscosity of the crust.

On Callisto, high-resolution maps of sections of Valhalla and the other large impact structures will yield new information concerning the nature of the concentric faulting. Possibly they will confirm the currently held idea that the faults formed later, after impact, in response to stresses generated when the material flowed back into the impact site.

Thus significant information that will contribute to a better understanding of the internal structure and ongoing surface processes can be obtained for each of these satellites. An understanding of the mutual influence that these bodies and Jupiter have had on each other as the system evolved will be limited by the degree of completeness of the data set. The investigators will be handicapped by a lack of information about how these satellites formed and an incomplete understanding of the physical processes that have molded them.

New Information on the Rings

Because a major goal of the mission is to obtain high-resolution images of the Galilean satellites, the spacecraft will orbit in the equatorial plane, as the satellites and rings do. Thus the instruments have an edge-on view of the ring, limiting observations to determine three-dimensional structure. While the *Galileo* instruments will measure the scattering properties in the infrared and provide more information about the particle distributions within the ring, the viewing-angle constraints, combined with the reduced data rate, will severely limit the observations. Still, the investigators are returning to a known system. This and the two-year duration of the mission will allow them to make specific measurements of the rings that, along with data concerning the magnetic field, will provide additional information related to their history.

Observations of the Magnetic Field

Constraints related to orbital insertion at the time for which the mission was originally planned required that the spacecraft spend a major portion of its time on the dark side of the planet. This did not delight the imaging team, but it allowed detailed sampling with the fields and particles instruments in the tail petal, the portion of the field that is swept outward, away from the sun. As the mission was delayed, constraints imposed by the geometry of approach required that the spacecraft be inserted into an orbit with even more time in the tail petal. If the high-gain antenna were fully functional this would have produced a large time-dependent data set. However, even then the constraints to prevent radiation damage would have ruled out detailed monitoring of the inner magnetosphere. Therefore the fields and particles people had not expected to gain much additional information concerning the interaction of Io and the inner satellites with the magnetic field. Instead, they expected their main yield to be related to temporal variations of the outer magnetosphere. Although the total amount of data will be sharply curtailed, significant results are expected.

Observations of the Planet

Even though the size of the high-resolution data set will be severely limited, it will certainly contribute to determining, to whatever extent possible, how the wind fields and cloud structures vary with time. The ability to observe at longer wavelength will reveal the spatial dependence of absorption by methane gas above the cloud deck. This will provide limited high-resolution data concerning the vertical structure of the clouds that was not available during the Voyager encounters.

Although some information concerning temporal variations in vertical motions and heat loss will be obtained, the unique observations that will be acquired by this mission will come from the atmospheric probe described above. When the probe is released, it will enter the north equatorial region and sample pressure, temperature and composition of the clouds as it descends through the atmosphere.

Careful analysis of the spacecraft signal during descent will yield clues about wind speeds at depths below the overlying cloud deck. After the probe ceases to transmit, the investigators plan to monitor the region into which it descended, along with other regions at the same latitude, to further characterize the area it sampled. The crucial data from the probe will be stored on the mother ship and sent home after the spacecraft has been inserted into orbit. Even if the unthinkable occurs and the spacecraft does not attain orbit, the engineers should be able to command the spacecraft to send this unique data set back to earth as the craft sweeps on by the planet.

If all goes well, the mission could extend for several years, and the fact that the spacecraft spends a great deal of time far from the satellites and on the dark side of the planet would allow us to slowly send home stored data. But an earlier decision, based on budgetary considerations, which led to the elimination of a wide-angle camera, has returned to haunt the mission. With the current data-rate constraints, such a camera would have been extremely valuable to get the bigger picture when the spacecraft is so close to Jupiter or a satellite. It would have complemented the tight view of the narrow-angle camera. Nevertheless, while the Voyager encounters stressed global coverage and helped to define the global circulation of the atmosphere and large-scale geology of the satellites, this mission can yield specific information that will help us to understand the dynamics or structure of selected regions of the atmosphere and satellites. The Galileo Mission will assist planetary scientists to better define the vertical structure of the atmosphere and the physical processes that have sculpted the Galilean satellites.

Additional Future Opportunities

This review of the advantages and limitations of the Galileo Mission illustrates the need for long-term global coverage of the planet as well as the monitoring of volcanic activity on Io. Long-lived spacecraft in orbit about Jupiter would be ideal; however none are planned for the near future. Earth-orbiting telescopes would allow observations at infrared and ultraviolet wavelengths and yield consistently higher resolution data than is readily attainable from the ground. The planetary camera on board the Hubble Space Telescope was designed to obtain observations of jovian cloud systems as small as 300 km in diameter and to allow multicolor imaging which could be coupled with ultraviolet spectra of selected regions of the planet for fifteen years (more than a jovian year). Even though original mirror problems degraded the early images, deconvolution techniques (using Fourier transforms) have yielded images that resolve features that are about 600 km in diameter. The new camera, installed in December 1993, has restored the original resolution and a time sequence of imaging and spectroscopic observations has begun even though competition with other astronomical objects limits Jupiter observations to less than 50 hours a year.

The Hubble Space Telescope (HST) was designed to have a long life. It was planned to evolve and serve as a workhorse to provide support for programs such as studying the time-dependence of Jupiter's atmosphere. A replacement for the Wide Field/Planetary Camera was being built and its installation date planned before the launch of HST and the discovery of the mirror problem. Modifications of this camera compensate for the mismatch of the telescope mirrors and allow higher resolution imaging. The first refurbishment mission will be followed by others. A new infrared instrument is under development for the second mission. The Hubble Space Telescope data surpasses ground-based data, but the demand for observing time is large. In addition, the operation and maintenance of any low earth-orbiting facility is difficult. Alternate approaches for observing the planets should therefore be considered.

Several years ago German and NASA scientists carried out a feasibility study for a 1-meter telescope in a 24-hour Earth orbit. Unlike the Hubble Telescope, this instrument would have been designed for optimal performance on bright objects, with a major stress on planetary observations. It could have been launched into a low earth orbit with a Delta 2 vehicle and then boosted into a 24-hour orbit with a second-stage booster. The 24-hour orbit, compared to the HST's 1.5-hour orbit, would allow data to be transmitted directly to a single ground-based station for a specified period each day and could eliminate the rapid heating and cooling cycle that plagues Hubble. Unlike the HST, the shuttle could not be used to service a craft in this high orbit. The telescope would be launched, used extensively for a limited time, and then turned off and ungraciously ignored as it continued in its long-lived orbit. Even though budgetary factors terminated this particular mission in the early planning stage, similar missions deserve consideration in the future.

While we are currently suffering from a dearth of spaceflight opportunities, ground-based facilities are improving. Recent advances in infrared cameras have greatly enhanced our ability to obtain spatially resolved multicolor wavelengths in this region of the spectrum where Jupiter radiates its own energy. In addition, the introduction of lightweight mirrors and improved support structures has led to construction of ever larger telescopes. Both honeycomb mirrors and multimirror laser-adjusted systems are being built. Parallel to the construction of these systems, progress is being made in developing techniques for measuring the smearing due to the earth's atmosphere in a closed-circuit system that allows the designers to optically compensate for the smearing with a deformable secondary mirror.

Meanwhile it seems that even though these new techniques have not reached maturity, Jupiter has decided it is time to challenge planetary observers by generating a spectacular event. In chapter 17 we will consider the planet's response to cometary collisions and how that may affect the Galileo Mission.

♃ Chapter 17

The Collision

On March 22, 1993, Eugene and Carolyn Shoemaker and David Levy discovered a strangely elongated object in the vicinity of Jupiter. Subsequent observations with telescopic systems capable of higher spatial resolution revealed that the object was a string of small bodies traveling through space. The most likely explanation was that a comet, in its orbit around the sun, had passed close enough to Jupiter that the resulting tidal forces had torn it apart. Normally comets that far from the sun are outgassing so slowly that they are not surrounded by a large gaseous coma and do not reflect enough sunlight to be seen readily. But because a dark insulating debris is left behind on the surface as the ices melt, when a comet nucleus fragments, not only is considerable dust generated, but fresh ice is also exposed. As the new surface outgasses, the comet brightens.

Such disruptions are rare. Occasionally a comet that is traveling in a very elongated orbit fragments during its close approach to the sun, but a similar event caused by Jupiter has been seen only once before the discovery of Shoemaker-Levy 9 (the ninth comet discovered by this team). In 1889 Comet Brooks 2, a comet-like object that was composed of eight bright regions, was discovered. Although this was three years before the earliest photographic observations, careful measurements of the object's path among the stars were recorded and an orbit was calculated. The astronomers concluded that three years earlier a comet traveling in an elongated orbit about the sun had made a grazing pass by Jupiter and that the tidal forces had disrupted the nucleus. The path of the comet was drastically altered, constraining it to a much smaller elongated solar orbit. In the months that followed

Figure 48. Comet Shoemaker-Levy. The cometary fragments are aligned along the orbit (the distinct line across the image) and a diffuse fan of dust is swept outward by the sun's radiation. Our view is from below the plane of the orbit. (Courtesy of J. Scotti, University of Arizona)

the Brooks 2 discovery, seven of the pieces, presumably flakes, evaporated. The surviving fragment still remains in a seven-year orbit, swinging out to Jupiter's distance from the sun and then returning to pass by the sun.

When modern observational techniques were brought to bear on Shoemaker-Levy 9, more that twenty fragments were reported. Its path among the stars was determined and its orbit computed. Brian Marsden, at the Smithsonian Astrophysical Observatory, reported that the comet had passed within 1.6 jovian radii from the planet's center in July 1992 and had become trapped in an orbit about Jupiter.

This is not an unknown occurrence. Studies of the orbits of several comets, including Gehrels 3, Oterma, and Helin-Roman-Crockett, indicate that they too have been in temporary orbits about Jupiter. This sort of entrapment is most likely for comets in nearly circular orbits with radii similar to Jupiter's orbital radius. As detailed studies of the near-Jupiter objects continue, the number of faint comets in Jupiter's family increases. Gene Shoemaker estimates that the rate of temporary capture of comets with diameters less than 2 km is about one per century. But the probability that one would pass near enough to Jupiter to be torn apart is much smaller.

As a larger arc of Shoemaker-Levy's orbit was obtained, Donald Yeomans, Paul Chodas, and Zdenek Sekanina, at the Jet Propulsion Laboratory, calculated an orbit that predicted that as the comet approached Jupiter in July 1994, it would pass closer to the center of Jupiter than one planetary radius. In other words, they were saying that the comet, approaching from the sunward side of the planet, would pass under the south polar region and smash into the planet at about 42° south latitude.

Available Energy

The popular press has concerned itself with the fact that an impact of a sizable asteroid would have devastating effects on the modern world and that an ancient asteroid impact led to the total extinction of the dinosaurs. One of the main reasons that such an event could have so much effect is that if a large metallic asteroid crashed into the ocean, it could expel a considerable amount of water into an orbit about the earth. The subsequent formation of a highly reflective ice cloud would cut off the sunlight, introduce huge changes in the earth's weather, and drastically affect all living creatures.

This raises a similar question concerning how Jupiter would respond to such an event. In a quest for an answer, a series of related questions should be asked. How massive is the comet and how fast is it traveling? Can it insert its energy into Jupiter's atmosphere or would it explode at high altitudes? If it does efficiently inject its energy into the atmosphere, how does this energy compare with the energy associated with the large jovian weather systems? If supersonic shocks develop, how would the atmosphere react? The answers to these and other, similar questions have been seriously pursued.

It is well known that the kinetic energy associated with a moving body is $1/2mv^2$, and the comet's impact velocity has been estimated as about 60 km/s. But an accurate determination of the mass of the comet has not been readily attainable. The manner in which the comet broke up has been utilized to set some limits, but the cohesive strength and rotation rate of the original nucleus are unknown. How the mass was distributed among the fragments is not known, either. At least three of the original flakes disappeared in 1993.

A "best estimate" is that the original body was composed of dirty ice with a density of about 1.3 gm/cc and a diameter of less than 2 km. This yields an estimate of the mass of less than 5.4×10^{12} kgm (6 trillion tons) which, moving at 60 km/s (130,000 mph), would carry a tremendous amount of kinetic energy. But remember that the cloud systems on Jupiter are huge and gravitational forces tend to crush the gases, causing rapid compression with depth.

The Red Spot spans about 26,000 km in longitude and about half that in latitude. The visible clouds form at a level where the gas pressure is about 0.6 times that at sea level on earth. The outer region of the spot rotates about the center at a speed of 100 m/s. The depth to which these winds extend is not known, but even if they only reach to a depth of 10 km (commercial planes fly at about 10 km, or 33,000 ft, to avoid our clouds), this giant system would contain as much energy as the impact of the "best guess" comet.

This estimate indicates that if 100 percent of the impact energy could be converted into rotational energy, a spot, probably smaller than the Red Spot and extending no deeper than 10 km, might form. But the question of where and how the energy is deposited in the atmosphere must be considered.

Off-Stage Activity

Orbital calculations indicated that the cometary fragments would collide with the planet at about 42° south latitude and about 20° of longitude beyond the visible edge of the planet. Therefore the planet would rotate more than a half hour before the points of impact would be illuminated and visible from the earth. Estimates made from data obtained in the fall of 1993 and spring of 1994 indicated that the impacts of the individual surviving fragments would span about a week, centered on July 19, 1994. Thus an earthbound observer would view the accumulated effect generated by the sequence of impacts.

The line of sight of the Hubble Space Telescope, in low earth orbit, is similar. Unhampered by turbulence in the earth's atmosphere and newly refurbished in December 1993, the planetary mode of the Wide Field Planetary Camera 2 obtained UV, visible, and IR images with more than 800 pixels, or samples, across Jupiter's 134,000 km polar diameter. This capability is highly useful for studying the atmospheric response to the collisions. In addition, the two UV spectrographs are more sensitive than any we have previously flown in space.

A desire to observe the points of impact led engineers at the Jet Propulsion Laboratory to use *Galileo* and Voyager to observe the planet. Still sixteen months away from encounter, the line of sight of the *Galileo* camera allowed it to see farther around into the dark side than earth-based cameras. In these images, the point of impact is on the dark limb of the planet. However, the diameter of the planet spans 65 pixels. This means that each pixel encompasses a region on the planet about 2000 km on a side, and the whole explosion is observed as a single bright pixel. This produces a good estimate of the energy released on impact but yields little information about how the explosion expands.

Reactivating the Voyager 2 camera or using its still-functioning UV spectrograph was also considered. The Voyager 2 spacecraft would be 43.4 AU (1 AU = Sun-Earth distance) from the sun. Looking back across 41.1 AU at Jupiter, Voyager would see the sun and Jupiter separated in the sky by less than 7°, and the entire planet would span 2.5 pixels in the narrow-angle camera. This did not justify reactivating the camera when the UV spectrograph could be programmed to acquire and store a set of observations. After the period of collisional activity was over, these data, consisting of linear arrays of brightness versus wavelength, could be sent to earth. Not only could these data be used to help establish times of impact, but the magnitude of brightening could be used to determine which events were most energetic.

Images obtained with the *Galileo* camera were stored, and the Voyager spectra were scheduled to be used to determine which of the two-dimensional arrays containing the images were given highest priority for read-back to the earth. In addition, careful ground-based observations of Io and Europa when they were on the far side were obtained to time the events. Reflection of the flash of the fireball revealed when the explosive impacts occurred.

The Significance of the Impact

The big question that investigators asked themselves was how the energy would be deposited in Jupiter's atmosphere. If the disturbance generated by the collision could excavate material down to a pressure level of about 6–10 atmospheres, water and ammonia gas could be carried up to the upper troposphere or even higher into the stratosphere. If considerable ammonia and water ice formed in the stratosphere, it would remain there for months or years. An extreme possibility was that a homogeneous stratospheric cloud could form, obscuring all cloud structure in the troposphere, creating a Uranus-like appearance. Although this would not have pleased those of us who devote our time to study the energy balance in the troposphere, it would have created a unique opportunity to determine how solar UV radiation alters the atmospheric ices. Atmospheric scientists would then have tried to determine how this high cloud changed the heat distribution as a function of altitude and how the atmo-

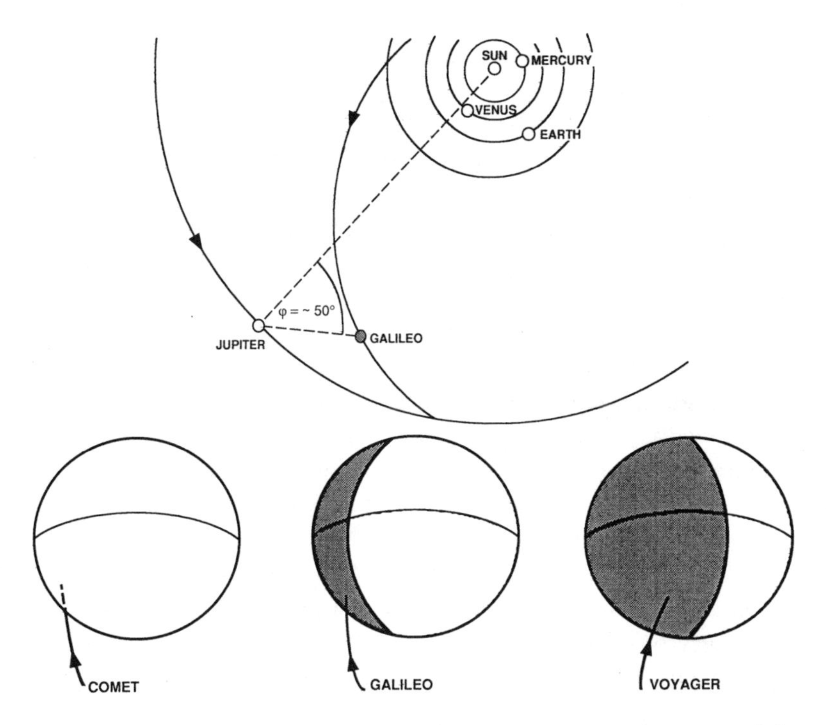

Figure 49. Positions of the Sun, Earth, and Jupiter at the Time of the Impact. The inserts show the viewing geometry of the Voyager and Galileo spacecraft compared with that of earth-based observers and Hubble Space Telescope (left). (JPL, NASA)

sphere responded to such a perturbation. Thus if the atmosphere had assumed a bland face, it would have been subjected to extreme scrutiny.

A lesser response was considered, where if the upwelling material were limited to the upper troposphere, vertical mixing and atmospheric waves would generate cloud systems that would serve as markers to indicate how the atmosphere responded to the collision. In either case, if the shock penetrated deep enough to excavate water, it would react with the molecules of ammonia and methane and generate significant changes in the UV spectrum of the atmosphere. Spectroscopists used the UV spectrographs on the Hubble Space Telescope and are working to document these changes.

A more violent response was also possible. This was one where the incoming cometary fragment would generate a supersonic shock and a fireball would form high in the atmosphere. Sprays of atmospheric gases and shards of the cometary fragment would be blown free of the planet. The mixture would go into orbit around Jupiter, where incoming UV sunlight would ionize the gases. The ions would then become trapped by the magnetic field. The force generated by the interaction of the charged particles and the rotating magnetic field would accelerate these particles poleward. As the particles spiraled into the polar region, they would be accelerated to high speeds. These speeding ions would collide with the atmospheric atoms and molecules and aurorae would be produced. Much of the interesting auroral phenomena occurs in the UV region of the spectrum.

With these various possibilities, it is apparent that competition for use of the HST was fierce. Administrators at the Space Telescope Science Institute selected a group of researchers and an observing program was prepared.

An additional effect of massive ejection of material from Jupiter is that material could settle into a tilted ring about Jupiter. Such a tilted ring would orbit the center of Jupiter in the plane of the comet's orbit; thus the particles would pass through the equatorial plane twice an orbit. The equatorial bulge of the planet would eventually pull these particles down into the equatorial plane, but that would take centuries. This means that *Galileo* may find a hazardous environment that contains extra debris. Hence the poor woe-ridden Galileo team will face still another obstacle as they attempt to carry out their mission.

Justification for My Observing Plans

The kinetic energy associated with billions of tons of ice plunging into the atmosphere at hundreds of thousands of miles per hour seems horrendous. But Jupiter is huge, and the fact that the pressure and density increase exponentially with depth makes it certain that the comet would suffer extreme braking as it entered the atmosphere. If the comet could dive unimpeded into the atmosphere, it would experience an increase in pressure (force per unit area) from 0.1 to 6 atmospheres of pressure (about 1 to 100 lb/in^2) in about 100 km, and would pass through the ammonia cloud deck into the lower water cloud deck. If it continued another 100 km, it would encounter pressures 300 times those at sea level on the earth (or about 4400 lb/in^2). In reality, the resistance that the comet fragments encountered generated extreme heating, which led to their rapid vaporization and to steam explosions. Just how deep the explosions penetrated, how much of the impact energy exploded back out into space and radiated away as fireballs, how much atmosphere was carried with it, and what the observations mean are the questions that we investigators are still asking.

My personal interest in this event was centered on how the atmosphere responded, and my efforts included obtaining high-resolution HST and ground-based images in visible and infrared light. After all the evidence has been sifted, we will have a reasonable estimate of the mass of the comet. That will allow us to estimate the depth of penetration and amount of energy that was inserted into the atmosphere. Although we have seen eruptive cloud deck phenomena in the past, such as the huge "thunderheads" associated with the South Equatorial Belt Disturbance during April and May 1993, we know neither the depth at which the disturbance originated nor the amount of energy that has dissipated in the lower atmosphere. Here these fragments can be considered depth charges and we have better knowledge of the associated energies.

It is useful to look at the sequence of events in considering the effect of these collisions. Twenty or more cometary pieces entered the atmosphere over a period of 7 days. If they were equally spaced, one fragment would have impacted the planet every 7 hours or so; but during that time the planet would have rotated about 250°, causing the impact sites to be well spaced in longitude. In Jupiter's hydrogen-rich atmosphere, where temperatures are on the order of 150°K, the speed of sound is about 500 m/s. The latitudinal spacing of the belt-zone cloud structure is 3–4° (3000–4000 km). Therefore, a pressure wave generated at the center of a zone and traveling at the speed of sound would require about an hour to expand to the edges of its zone. Because the zonal winds are small (30–40 m/sec), the disturbance should continue to spread beyond the zone.

In an effort to determine how the atmosphere would react and whether anticyclonic storms would form, Timothy Dowling, at Massachusetts Institute of Technology, computed the effect of a single impact (see plate 8). His general circulation model predicted that the wave would spread outward, crossing the belt-zone interfaces as a pebble would disturb a pond. Dowling's simplified model did not address the initial impact; instead he assumed that the comet would reach 5 bars of pressure and studied how such an energy surge would perturb the pressure at that level.

Mark Marley, at New Mexico State University, considered another possibility. He proposed that the steam explosion would generate a quake or seismic waves, which would propagate downward. If density discontinuities were present in the planet's interior, the seismic waves would be reflected upward and emerge as a coherent ring where the pressures could vary enough to cause whiter clouds to form. This phenomenon could be separated from Dowling's waves because the response time is about an hour while Dowling's waves travel more slowly and would not be readily detectable for several hours.

The question of how the energy would be deposited has been addressed by various groups that have traditionally been associated with defense projects and have recently adapted their "bomb codes" to study the effect on the atmosphere if an asteroid were to collide with the earth. Several of them have modified their programs to study this problem. They predicted that because the entry speed of the comet was more than 100 times the local speed of sound, strong supersonic shocks would develop. The way the cometary particle came apart and the manner in which the energy was dispersed would determine where the energy was deposited in the atmosphere.

A typical example of this type of calculation is illustrated in figure 50. David Crawford and Timothy Trucano, from Sandia Laboratories, predicted that the comet would penetrate to a depth of 140 km in less than 2.5 seconds after it entered the atmosphere. In this model temperatures as high as tens of thou-

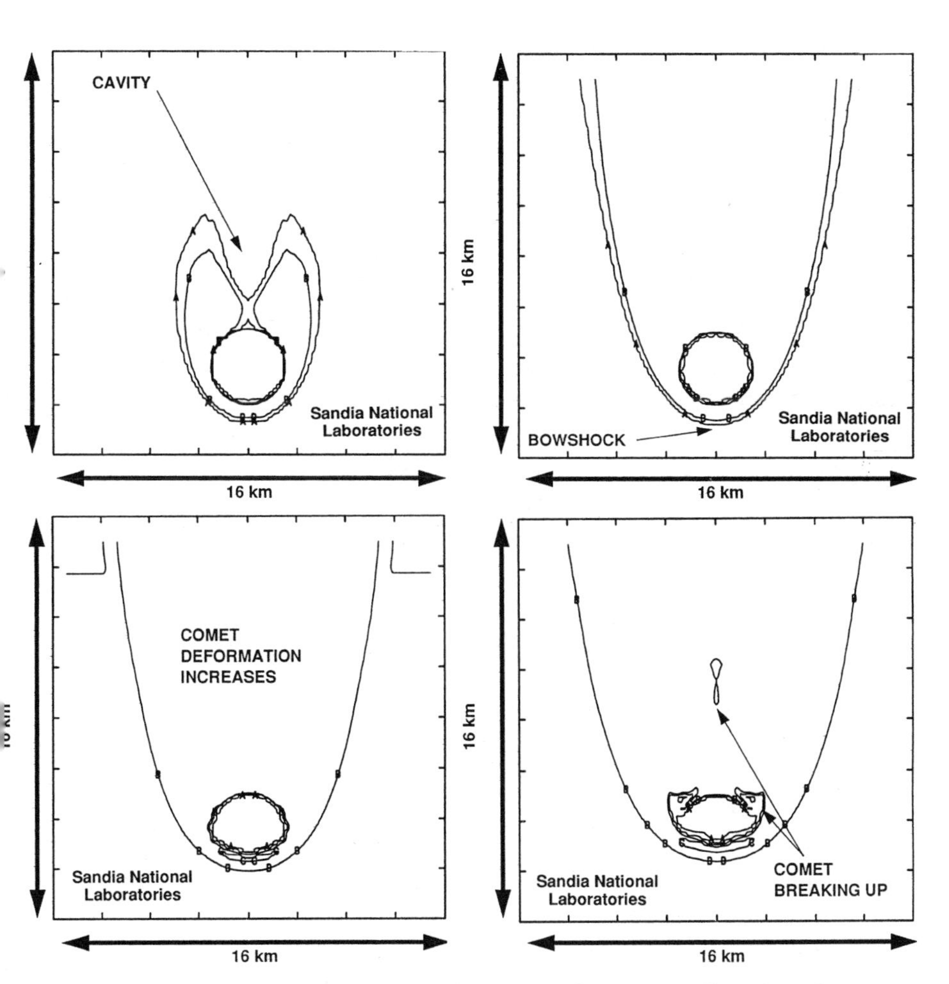

Figure 50. A Comet Impact Model. This series of diagrams illustrates the development of the shock as a sphere of ice 3 km in diameter encounters Jupiter's atmosphere. The contour lines A, B, and C indicate temperatures of 400°, 4000°, and 40,000°K. Left to right, the top diagrams are at 0.1 and 1.0 sec and the bottom diagrams are at 2.0 and 2.4 sec after entrance of the comet. (Courtesy of T. Trucano, and D. Crawford, Sandia Laboratories)

sands of degrees were predicted. This model did not deal with how the atmosphere would react to such extreme heating but concentrated on the stress imposed on the comet.

Varying results from collisional models indicated that the effective depth in the atmosphere where the energy was inserted is extremely uncertain, but some basic relations must hold true. If a local region were heated to 15,000°K (27,000°F), it would radiate heat away 100 million times faster than the cold cloud deck and generate an impressive flash. It would cool rapidly in the early stages; but as the spot cooled, it would lose energy more and more slowly (see appendix 6). If minimal mixing with its local surroundings occurred, it could be weeks or months before the spot totally lost its identity. Such a hot spot could act as an obstacle to the local wind flow. Because the atmosphere of Jupiter is at a temperature where ammonia readily freezes or melts, the induced perturbations should generate cloud patterns that will serve as markers to indicate how the atmosphere is responding to the remaining hot spots. It is this atmospheric response that I have been studying. This requires observations before, during, and after the event.

Summary

We anticipated a series of collisions that would be longitudinally distributed. Just how deep the energy would be inserted or how violent the resulting fireballs would remain to be seen. The planet-comet interaction could have resulted in a range of atmospheric responses that could lead to a homogeneous white cloud deck or to the establishment of interesting cloud systems, possibly even large closed eddies like the Red Spot and White Ovals. Aurorae and tilted dust rings might also be formed.

Personally, I had hoped for modest "salting" of the stratosphere and for lingering hot spots in the troposphere. Scattered condensations of white ice in the stratosphere would allow us to detect poleward motion and to determine the global pattern of circulation in the stratosphere. Although previous to the comet impact we had evidence of equatorial upwelling and expected a descending polar flow, we had not detected it.

Tropospheric hot spots would interact with the zonal winds, creating trains of responding wave patterns. The nature and speed of these waves could be compared with those present in the Voyager images during an undisturbed era to determine how the atmosphere responds to such an event.

I did not expect a new red spot, nor even a large white oval. This was because I anticipated that very little of the impacting energy could be converted to organized motion within this type of long-lived weather system. The historical data related to the 1938 formation of the White Ovals indicate these eddies were formed when a convective blob carried heat from Jupiter's interior into the region below the visible cloud deck. In these regions the densities are so great that a temperature increase of a few degrees could have inserted a great deal of energy into the upper troposphere, where these weather systems are seen. As this book goes to press, the planet's response is still unknown, so if a new giant eddy forms, I will say "please pass the crow." In the meantime I have prepared to observe interesting atmospheric wave phenomena and seek a chance to gain a better understanding of the global circulation of this giant planet.

Epilogue

Jupiter, the most accessible of the outer planets, will continue to fascinate us. The cometary collision and the Galileo Mission will provide continuing stimulus for the coming years, but it is not clear what the next generation of spacecraft will be.

Among the astronomical community there are many individuals who dream of specialized jovian orbiters committed to a significant but limited set of observations, and of a battery of probes to learn more of the conditions below the opaque cloud deck. These might be small systems that would relay data to an orbiting communicator, a specialized ship committed to receiving and sending data. If no short-lived special mission were under way, the communication station could be placed in idle mode in a parking orbit until it was again needed.

Others dream of large lunar-based observatories. Here on this windless, slowly rotating satellite, with the telescope shaded from the sun by a large, lightweight screen and no obscuring atmosphere, astronomers could observe Jupiter continuously for more than two weeks every month. They would watch the giant planet rotate before the camera and could sample developing cloud systems several times during the four to five hours that the cloud would be readily visible on the lunar-facing side of Jupiter. Every ten hours they could repeat the observing scheme. With a large thermally stable telescope and remote cameras that were operated by a staff working in a multipurpose lunar station or remotely from the earth, investigators could sample the development of cloud systems or storms on the giant alien planet. A remote possibility? Certainly! Nevertheless I am sure that we will see equally far-fetched schemes come to fruition and that the jovian system holds enough challenges that our descendants, the students of our students, will continue to wrestle with them.

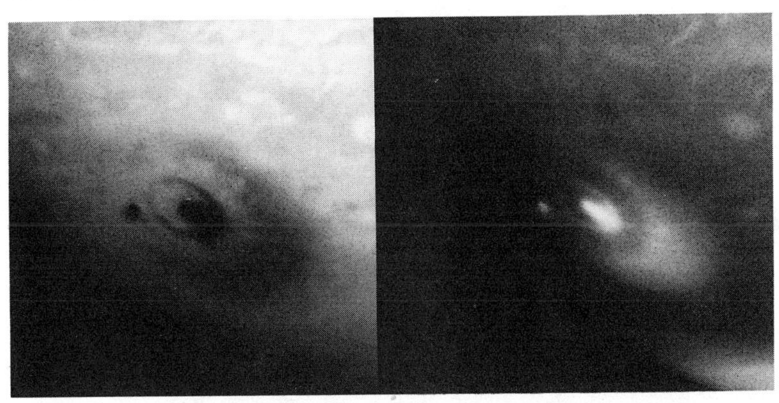

Hubble Space Telescope Images of Impact Sites D and G. The small site to the left is from comet fragment D (impact, July 17, 1994); on the right, the G site (impact, July 18) is one of the largest. These views show material that has been ejected from the G site trending toward the southeast: the high expanding plume appears transparent around its outer edges. Many of the features seen in these images are formed high in the stratosphere. The green filter, left, samples the ammonia cloud deck and the methane filter, right, samples high in the atmosphere. (Hubble Space Telescope Comet Team)

Comet Impact Sites C, A, and E. The view on the right shows these sites in ultraviolet light; on the left, the same regions as they appear in violet light. The Hubble Space Telescope is the only telescope capable of observing Jupiter in ultraviolet light at this resolution. Because ultraviolet light is absorbed by the earth's atmosphere, an orbiting telescope is necessary. (John Clarke, University of Michigan, and NASA)

♃ Appendix 1

Physical Parameters of Jupiter and the Satellites

The Physical Parameters of Jupiter Compared to Earth and the Other Giant Planets

Characteristic	*Earth*	*Jupiter*	*Saturn*	*Uranus*	*Neptune*
Mass (earth masses)	1.0	317.9	95.2	14.5	17.1
Equatorial Radius (R_e)	1.0	11.21	9.45	4.0	3.9
Ellipticity (e)[1]	.0034	.0649	.0980	.0229	.017
Mean density (gm/cc)	5.52	1.33	0.69	1.29	1.64
Rotation Period[2]	$23^h55.8^m$	$9^h55.5^m$	10^h39^m	17^h14^m	16^h07^m
Inclination of Equator (deg)	23.45	3.12	0.69	1.29	1.64
Mean Distance from Sun (AU)	1.00	5.20	9.54	19.19	30.06
Period of Revolution[3] (yr)	1.00	11.86	29.46	84.01	164.79
Orbital Eccentricity[4]	.0167	.0483	.0560	.0461	.0097
Inclination to Ecliptic[5] (deg)	0.000	1.308	2.488	0.774	1.774

1. Polar radius = $(1-e)R_e$
2. This is the earth's rotation period relative to the stars = one siderial or stellar day.
3. This is the length of the planetary year.
4. Ratio of the separation of the two focii of the elliptical orbit to the major axis.
5. The ecliptic is defined as the plane of the earth's orbit around the sun.

Adapted from *The New Solar System*. 3d ed. Ed. by J. K. Beatty and A. Chaikin. Cambridge, Mass.: Sky Publishing Corp., 1990.

The Physical Parameters of the Jovian Satellites

Satellite	*Year of Discovery*	*Mean distance from planet (km)*	*Period of Revolution (days)*	*Orbital inclination (deg)*	*Eccentricity*	*Radius (km)*
Metis	1979	127,960	0.295	0[1]	0[1]	20[1]
Andrastea	1979	128,980	0.298	0[1]	0[1]	12 × 8
Amalthea	1892	181,300	0.498	0.4	0.00	135 × 7
Thebe	1979	221,900	0.675	0.8[1]	0.01[1]	50[1]
Io	1610	421,600	1.769	0.4	0.00	1815
Europa	1610	670,900	3.551	0.47	0.01	1569
Ganymede	1610	1,070,000	7.155	0.19	0.00	2631
Callisto	1610	1,883,000	16.689	0.28	0.01	2400
Leda	1974	11,094,000	238.72	27	0.15	8[3]
Himalia	1904	11,480,000	250.57	28	0.16	90[3]
Lysithea	1938	11,720,000	259.22	29	0.11	20[3]
Elara	1905	11,737,000	259.65	28	0.21	40[3]
Ananke	1951	21,200,000	631	147[2]	0.17	15[3]
Carme	1938	22,600,000	692	163[2]	0.21	22[3]
Pasiphae	1908	23,500,000	735	147[2]	0.38	35[3]
Sinope	1914	23,700,000	758	153[2]	0.28	20[3]

1. Observed for the first time with Voyager. Values are uncertain.
2. Angle of inclination greater than 90° indicates retrograde revolution.
3. Estimated apparent radii, based on observed brightness and color to imply percent reflectivity.

24 Appendix 2

Color, Wavelength, Frequency, and Wave Number

Several units are in use to specify the color (wavelength or frequency) of light; the reasons for this are twofold. First, it is convenient to use small numbers to designate a quantity. Because the range of wavelengths associated with electromagnetic radiation spans orders of magnitude, several terms are used to indicate wavelength. Second, because the energy associated with a photon of light is proportional to the frequency or inversely proportional to the wavelength, it is also convenient to refer to the color of the radiation in frequency units.

The fundamental relationship between wavelength and frequency is illustrated in figure A-1.

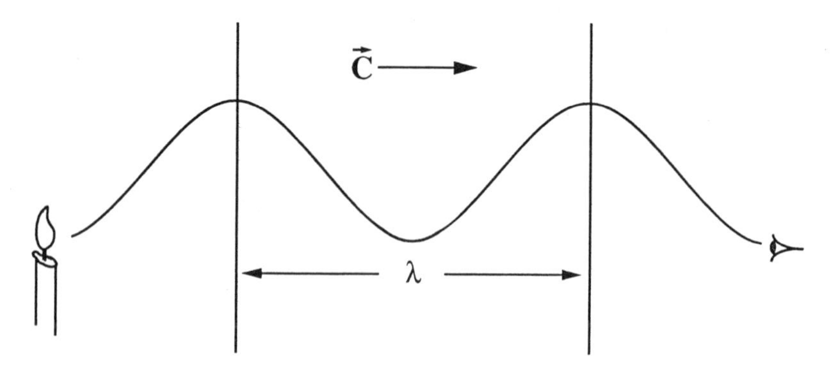

Figure A-1. The Relation Between Wavelength and Frequency. Here C is the speed of light, λ is the wavelength. As the light from the source arrives, the observer will see the amplitude fluctuate ν (or C/λ) times per second, where ν is the frequency.

The following table defines some of the units that are in current use.

Unit	*Relation to*	*Region of Spectrum Applied*
	Wavelength Units	
Angstrom	1/10,000,000,000 meters	visible, ultraviolet, X ray, and Gamma ray
Nanometers	1/1,000,000,000 meters (this is a standard metric unit)	visible, ultraviolet, X ray, and Gamma ray
Microns	1/1,000,000 meters	near infrared
Centimeters	1/100 meters	far infrared
Decimeters	1/10 meters	radio
Decameters	10 meters	radio
	Frequency and Wave Number Units	
Frequency*	cycles/second (Hertz) *(the speed of light)/* wavelength	all wavelengths
Wave number*	1/centimeters (1/wavelength — yields values in the hundreds and thousands in the infrared)	infrared

* Equal increments of frequency or wave number correspond to equal increments of energy.

2♃ Appendix 3

Systems of Longitude and Latitude

Standard Systems of Longitude

One result of the analysis of the data concerning the magnetic field of Jupiter was that the period of rotation of the magnetic field was accurately deduced. Although it is impossible to measure the rotation rate of the interior of the planet directly, it can be assumed that the magnetic field is imbedded in the conductive core and is co-rotating with it. Therefore, although the radio emission is generated in the upper atmosphere, or magnetosphere, the observed periodicity of the data reveals the rotation rate of the magnetic field and hence, of the core. If the rotation rate of the magnetic field is used as a basis for mapping the displacement measurements of cloud features in the visible cloud deck, the resulting wind velocities should convey information about the energy balance of the atmosphere relative to the core. Within this context, the accurate determination of the radio period based on the observed variability of the radio signal was significant to a large group of investigators other than those who were attempting to understand the nature of magnetic fields within the solar system.

When a new definition is needed, astronomers who are members of the International Astronomical Union (IAU) have traditionally established definitions and accompanying nomenclature. The IAU's work is carried out by subgroups, or commissions, with international membership. The commissions make specific recommendations to the general assembly, which meets every three years. Standards developed by the IAU have widespread acceptance within the international scientific community and provide a basis for worldwide exchange of data and information. Commission 40 was the appropiate working group to address the question of the standard period of rotation.

In 1962 a working group proposed a period of $9^h55^m29.37^s$, based on the recurring bursts of the decametric radiation. Previously the IAU had accepted two other rotation periods for Jupiter: System I, with a period of $9^h50^m30.0034^s$, and System II, with a period of $9^h55^m40.6322^s$. These periods had been derived from average translation rates near the equator for System I and near the Red Spot over an extended period of time for System II. Data for a given apparition, or season of viewing, could be plotted within these systems and the resulting plot of longitude versus time would generate a chart that would allow small accelerations of the features to be recognized as changes of the slope of the longitude versus time relationship. Although these systems of longitude provided a framework within which changes in the motions of clouds could readily be recognized, they did not provide a physical system within which the data could be interpreted as energy transport relative to the core of the planet.

Even though astronomers had accumulated about five years of data by 1962, its variability limited the accuracy to which the period could be determined. As time passed, it became apparent that the proposed period was too short. In 1975, after accumulating thirteen more years of data, radio astronomers increased the period by 0.34 seconds. Although this does not appear to be a large correction, it accounted for an accumulation of three degrees of longitudinal shift per year. In other words, when the data was plotted in the old system, the location of the large noise bursts drifted three degrees eastward each year; hence the error in the period was quite noticeable.

To define a unique longitudinal system on a rotating planet, the remote observer must not only know the rate of rotation of the planet but must also select some point in time at which the longitude coinciding with the central meridian of the planet is assigned a value. The first radio system was referenced to Jan. 1, 1957, and was designated System III(1957); while the system proposed in 1975 was referenced to Jan. 1, 1965, near the midpoint of the data, and was designated System III(1965). This system is currently the standard one in use and is defined in terms of the initial central longitude at 0^h UT, Jan. 1, 1965, and a rate of rotation of exactly 870.536° per Julian day, with longitude increasing westward, or as time progresses. In an effort to associate this system with previous systems of longitude, the longitude at 0^h UT, Jan. 1, 1965, was not set at zero, but was defined as 217.956°; hence the longitude in System III(1965) can be computed as follows:

$$L(t) = 217.956° + 870.536(t - t_0)$$

where L(t) is the longitude in question at a specific time and $t - t_0$ is the total time in days that has elapsed since 0^h UT, Jan. 1, 1965. This system is defined by A. C. Riddle and J. W. Warwick, *Icarus* 27 (1976), 457–59. Conversion from Systems I and II are as follows:

$$L_{III} = L_I + 35.601 - 7.364(t - t_0)$$

and:

$$L_{III} = L_{II} + 81.245 + 0.266(t - t_0)$$

where t_0 is 0^h UT on 1 Jan. 1965.

Planetocentric vs. Planetographic Latitude

To the uninitiated it would seem that the angle formed between the equatorial plane of the planet and a line connecting the center of the planet to the point of interest in the cloud deck would be the latitude. As we look at the atmosphere of Jupiter, however, the oblateness, or ellipticity, of the planet causes the polar region to drop away from our line of sight more rapidly than would be the case on a spherical planet. The clouds appear foreshortened and the angle of solar illumination is exaggerated. To correct for this, observers and modelers have assumed that the distorted shape is an ellipsoid that is defined by the ratio of its equatorial radius, R_e, to its polar radius, R_p. They have used a latitudinal system that is referenced to a vector that is directed along the local vertical. The combined effects of rapid rotation and gravitational force act downward along this line and the clouds form perpendicular to this reference vector. This latitude is called the planetographic latitude, while the latitude referenced to the center of the planet

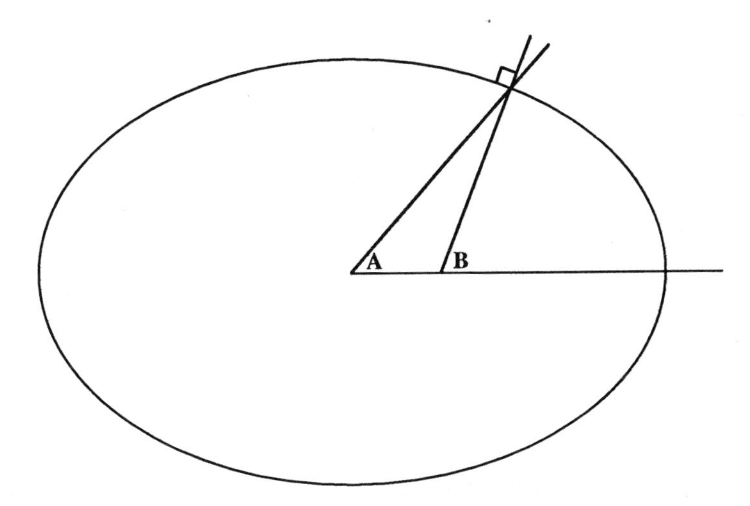

Figure A-2. Planetocentric and Planetographic Latitude. A is planetocentric and B is planetographic.

is called the planetocentric latitude. The planetographic latitude is the natural coordinate of local events at the level of the cloud deck. This latitude is equivalent to the latitude that a surveyor would derive at some point of the earth. On the other hand, the planetocentric latitude is unambiguous at all distances from the planet center and is the obvious coordinate for problems dealing with the interior or magnetosphere.

Because the Greek counterpart to Jupiter is Zeus, you may also find these jovian latitudinal systems referred to as the zenographic and zenocentric latitudes. The conversion from planetocentric to planetographic latitude is

$$\tan(\mathrm{lat}_g) = (R_e/R_p)^2 \tan(\mathrm{lat}_c)$$

where R_e/R_p is the ratio of the equatorial radius to the polar radius and lat_c and lat_g are the planetocentric and planetographic latitudes, respectively. On an oblate planet the two latitudinal systems are equal at the equator and the pole and differ most widely at mid-latitudes. In the case of Jupiter the difference is as much as 3.5°, with the north edge of the North Tropical Zone located at about 20° zenocentric latitude and at about 23° zenographic latitude.

24 Appendix 4

Temperature Scales

There are three temperature scales that are commonly used: the Fahrenheit scale, which is still used for everyday applications in the United States; the Celsius, or centigrade, scale, which is used in much of the world; and the Kelvin scale, which is used when we are interested in the total energy content of a system. The following relationships allow us to convert among these three scales.

$$K = {}^\circ C + 273$$
$${}^\circ C = 5/9 \times ({}^\circ F - 32)$$
$${}^\circ F = 9/5 \times {}^\circ C + 32$$

The following scale may help you with the conversion.

	Fahrenheit	*Celsius*	*Kelvin*
Absolute Zero	−459.67	−273.15	0.00
Freezing Point of Water	32	0	273
Boiling Point of Water	212	100	373

24 Appendix 5

Operation of an Interferometer

The Voyager infrared spectrometer was an interferometer. To understand how an interferometer works, consider a laser beam composed of a single color of light. As the beam travels along the optical path from the collecting optics to the detector, the frequency with which the associated electric and magnetic fields vary will be determined by the wavelength of the laser that we have chosen. The number of fluctuations along the path will depend on the length of the path, and intensity measured by the detector will be related to the amplitude of the fields. An interferometer takes advantage of these properties of light. In the interferometer the incoming beam of light encounters a partially silvered diagonal mirror. Part of the light passes through and part is reflected to the side. The two components of the split beam then encounter reflecting mirrors which return the components to the beam splitter. Part of the reflected beam passes through and part of the transmitted beam is reflected (see figure A-3). Although a fraction of the light will be scattered and transmitted out of the desired path, components of the original beam will be recombined. As long as the two independent paths are exactly the same length or vary by some exact number of wavelengths, the electric and magnetic fields will arrive at the detector in phase and add constructively. The only noticeable effect will be a reduction in the intensity due to light losses within the instrument. However, if the path lengths vary by exactly some number plus one-half a wavelength, the fluctuations of the fields will be completely out of phase and they will combine destructively, canceling each other and causing the detector to record a loss of light. Because of this, if we design the interferometer so

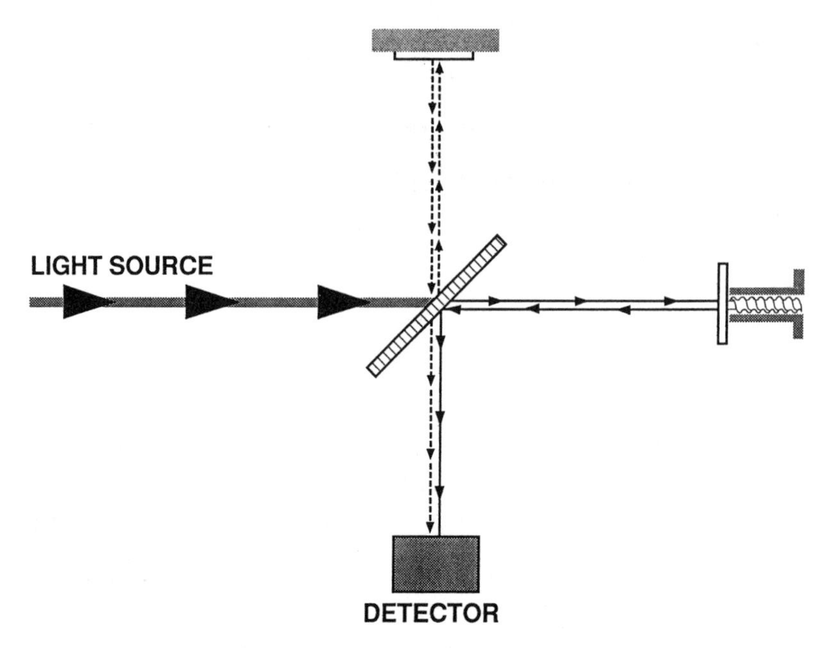

Figure A-3. A Simple Interferometer. The incoming beam of light (dark arrows) is split into two components which follow different paths to recombine before reaching the detector. The dotted line traces the path of the component that is scattered by the partially silvered diagonal mirror and reflected from the fixed mirror, and the solid line traces the transmitted component that is reflected from the adjustable mirror.

that one of the reflecting mirrors is precisely driven and the intensity of the light is sampled rapidly relative to the driving rate, the detector will record fringes, or systematic variation, in the intensity of the signal. If our driving screw is precisely calibrated, the displacement of the screw needed to move from one fringe to the next will be one-half the wavelength of our laser (the beam travels out and back).

This simple idea was utilized for the Voyager interferometer, but all wavelengths of light were measured simultaneously. It is apparent that this is the natural approach and that it would be efficient to measure all colors of light simultaneously. Even though it seems that this would create irretrievable mixing of the signals from the different wavelengths, this is not the case. If the mirror is steadily displaced over an adequate distance and the variation in the intensity of the signal is rapidly and evenly sampled as a function of time, a numerical operation called a Fourier transform can be performed on the data. Unbelievable as it seems, the spectrum, a map of how the light varies with color, can be recovered.

♃ Appendix 6

Radiation from a Warm Body

The rate of energy loss is a strong function of the temperature of the exposed surface of a body. The physical basis for understanding the rate of heat loss from any kind of object, including large planets, was formulated early in this century. This analysis was based on heat loss as a function of absolute temperature where the zero point is set at 0° Kelvin (−273°C or −459°F), the temperature at which all translational molecular motion ceases (see appendix 4). Max Planck (1858–1947) developed a statistical argument and generated a mathematical relationship that describes the wavelength, or color dependence, of the energy emitted per unit area by a body at a given temperature in *thermal equilibrium* with its surroundings. This relationship predicts that the rate of emission of the total energy at all wavelengths per unit area is proportional to the fourth power of temperature (T^4), and the derivative of the function can be used to show that the wavelength at which the maximum energy per unit area is emitted is inversely proportional to the temperature ($1/T$). These latter two results had been deduced from laboratory measurements by Josef Stefan and Ludwig Boltzman and by Wilhelm Wien. Therefore, if a body is in thermal equilibrium, the relation between temperature and both the wavelength distribution of the emitted radiation and its ability to radiate heat are well understood.

The amount of energy emitted by the sun also depends on color, or wavelength, and nearly fits a Planck function for a temperature of about 5800°K, emitting the maximum energy per unit area in the yellow region of the spectrum. If Jupiter is radiating its own heat, with an absolute cloud-deck temperature of 150°K (−123°C), it will radiate most of this energy in the infrared region of the spectrum. Therefore, to measure Jupiter's radiation field and to understand rate of heat loss, detailed infrared observations are required; but consideration of the total energy that the Planck function or the Stefan-Boltzman law would predict yields some insight into the problem. A temperature of 300°K (27°C or 80°F) would not be uncommon for a parcel of air in the earth's atmosphere; however, the rate of heat loss from such a parcel would be 16 (2^4) times greater than the rate from the 150°K ammonia-ice cloud deck in Jupiter's atmosphere. Although these fundamental relationships provide arguments for expecting low rates of heat loss and long adjustment times in Jupiter's atmosphere, the actual rate of heat loss from the interior is determined by the degree to which the upper atmosphere serves as an insulating blanket in the infrared and the extent to which vertical mixing dominates the energy balance of the observed cloud deck.

Additional Readings

Beatty, J. Kelly, and Andrew Chaikin. *The New Solar System.* Cambridge, Mass.: Sky Publishing Corporation, 1990.

Belton, Michael, Robert West, and Jurgen Rahe. *Time-Variable Phenomena in the Jovian System.* NASA Special Publication 494. Washington, D.C., 1989.

Fimmel, Richard, William Swindell, and Eric Burgess. *Pioneer Odyssey.* NASA Special Publication 396. Washington, D.C., 1977.

Greeley, Ronald. *Planetary Landscapes.* London: Allen & Unwin, 1985.

Hartmann, William. *Moons and Planets.* 3d ed. Belmont, Calif.: Wadsworth Publishing Company, 1993.

Hunt, Garry, and Patrick Moore. *Jupiter.* New York: Rand McNally & Co., 1981.

Morrison, David, and Jane Samz. *Voyage to Jupiter.* NASA Special Publication 439. Washington, D.C., 1980.

Peek, Bertrand. *The Planet Jupiter.* Rev. by Patrick Moore. London and Boston: Faber and Faber, 1981.

Smith, Bradford. "Voyage of the Century." *National Geographic* (August 1990), pp. 49–65.

Voyager Team. "Mission to Jupiter and Its Satellites: Voyager 1." *Science* (June 1, 1979), pp. 945–1008.

Voyager Team. "Mission to Jupiter and Its Satellites: Voyager 2," *Science* (November 23, 1979), pp. 925–995.

Periodicals

Astronomy (ISSN 0091–6358). Published monthly by Kalmbach Publishing Company, Waukesha, Wisc.

Mercury (ISSN 0047–6773). Published bimonthly by Astronomical Society of the Pacific, San Francisco, Calif. (Open to the general public.)

The Planetary Report (ISSN 0736–3680). Published bimonthly by The Planetary Society, Pasadena, Calif. Available to members of The Planetary Society. (Open to the general public.)

Sky and Telescope (ISSN 0037–6604). Published monthly by Sky Publishing Company, Cambridge, Mass.

Index

Numbers in italics indicate illustrations.